# Current Topics in Microbiology and Immunology

Volume 377




For further volumes:
http://www.springer.com/series/82

Harris G. Fienberg · Garry P. Nolan
Editors

# High-Dimensional Single Cell Analysis

## Mass Cytometry, Multi-parametric Flow Cytometry and Bioinformatic Techniques

Responsible Series Editor: Bernard Malissen

Springer

*Editors*
Harris G. Fienberg
Garry P. Nolan
Department of Microbiology
and Immunology
Center for Clinical Sciences Research
Stanford School of Medicine
Stanford, CA
USA

ISSN 0070-217X ISSN 2196-9965 (electronic)
ISBN 978-3-642-54826-0 ISBN 978-3-642-54827-7 (eBook)
DOI 10.1007/978-3-642-54827-7
Springer Heidelberg New York Dordrecht London

Library of Congress Control Number: 2014936870



Printed on acid-free paper

Springer is part of Springer Science+Business Media (www.springer.com)

*To CC:*
*Thanks for the pep talks and for the straight talks and for always knowing which one I needed. I love you.*

# Preface

High-dimensional cytometry has ushered in a new era of single-cell analysis. High-end flow cytometers are now capable of 15-parameter analysis and the mass cytometer is routinely used for 34 parameter experiments and is capable of analyzing 100 unique parameters on single cells.

These technological advances have enabled a comprehensive panel of surface markers to be analyzed in tandem with intracellular protein states, allowing researchers to disentangle complex signaling networks in heterogeneous tissues such as blood, bone marrow, and tumors in ways that were previously impossible. Mass cytometry and high-dimensional flow cytometry have been employed in transformative studies in diverse disciplines including hematopoiesis, immunology, and drug profiling. The rapid increase in dimensionality has also spurred the development of novel analytics allowing researchers to probe and visualize high-parameter, single-cell datasets.

In this volume we will address the most interesting questions and applications enabled by high-dimensional technologies, review current practical approaches used to perform high-dimensional experiments, and address the key bioinformatic techniques developed to facilitate analysis of datasets involving dozens of parameters in millions of single cells.

High-dimensional cytometry has made it possible to systematically measure mechanisms of tumor initiation, progression, and therapy resistance on millions of individual cells from human tumors. This has ushered in a "single-cell systems biology" view of cancer ("High-Dimensional Single-Cell Cancer Biology"). High-dimensional cytometry has facilitated a similar paradigm shift in immunology and provided a view of the human "immunome" with unprecedented breadth ("Studying the Human Immunome: The Complexity of Comprehensive Leukocyte Immunophenotyping") and allowed for the exploration of immunological cell types, such as $CD8^+$ T cells with increasing depth ("High-Dimensional Analysis of Human $CD8^+$ T Cell Phenotype, Function, and Antigen Specificity"). Mass cytometry has provided an increasingly sophisticated view of intracellular signaling and acted as an ideal tool to pry open the signaling processes of cancer ("Mass Cytometry to Decipher the Mechanism of Nongenetic Drug Resistance in Cancer").

New techniques have emerged to maximize the power of high-dimensional cytometry. Mass cell barcoding greatly increases the throughput, reduces antibody consumption, and increases data quality for mass cytometry experiments

("A Practical Guide to Multiplexed Mass Cytometry"). Proximity ligation assays greatly expand the number of possible processes that can be targeted, allowing for the detection of protein–protein interactions, post-translational modifications, and interactions of proteins with nucleic acids ("Analysis of Protein Interactions in situ by Proximity Ligation Assays").

A host of new analytical approaches and platforms have been developed to analyze increasingly complex high-dimensional single-cell datasets. Cytobank, an analysis platform leveraging recent advances in cloud computing and virtualization, lets researchers annotate, analyze, and share results along with the underlying single-cell data ("Cytobank: Providing an Analytics Platform for Community Cytometry Data Analysis and Collaboration"). Advances in unsupervised discovery allow for biological insights to be gleaned from large datasets without *a priori* knowledge or intensive manual intervention ("Computational Analysis of High-Dimensional Flow Cytometric Data for Diagnosis and Discovery"). In order to deal with intricate intracellular data, computational deconvolution approaches have been developed to reconstruct and describe signaling dynamics ("Shooting Movies of Signaling Network Dynamics with Multiparametric Cytometry").

Finally, looking into the future, new technologies such as hyperspectral cytometry may be poised to increase parameterization capabilities of single-cell measurement and expand the capabilities of high-dimensional cytometry ("Hyperspectral Cytometry").

Harris G. Fienberg
Garry P. Nolan

# Contents

# High-Dimensional Single-Cell Cancer Biology

**Jonathan M. Irish and Deon B. Doxie**

**Abstract** Cancer cells are distinguished from each other and from healthy cells by features that drive clonal evolution and therapy resistance. New advances in high-dimensional flow cytometry make it possible to systematically measure mechanisms of tumor initiation, progression, and therapy resistance on millions of cells from human tumors. Here we describe flow cytometry techniques that enable a "single-cell systems biology" view of cancer. High-dimensional techniques like mass cytometry enable multiplexed single-cell analysis of cell identity, clinical biomarkers, signaling network phospho-proteins, transcription factors, and functional readouts of proliferation, cell cycle status, and apoptosis. This capability pairs well with a signaling profiles approach that dissects mechanism by systematically perturbing and measuring many nodes in a signaling network. Single-cell approaches enable study of cellular heterogeneity of primary tissues and turn cell subsets into experimental controls or opportunities for new discovery. Rare populations of stem cells or therapy-resistant cancer cells can be identified and compared to other types of cells within the same sample. In the long term, these techniques will enable tracking of minimal residual disease (MRD) and disease progression. By better understanding biological systems that control development and cell–cell interactions in healthy and diseased contexts, we can learn to program cells to become therapeutic agents or target malignant signaling events to specifically kill cancer cells. Single-cell approaches that provide deep insight into cell signaling and fate decisions will be critical to optimizing the next generation of cancer treatments combining targeted approaches and immunotherapy.

J. M. Irish (✉) · D. B. Doxie
Vanderbilt University, Nashville, TN, USA
e-mail: jonathan.irish@vanderbilt.edu

Current Topics in Microbiology and Immunology (2014) 377: 1–21 
DOI: 10.1007/82_2014_367

Published Online: 27 March 2014

## Contents



# 1 Introduction

Single-cell approaches reveal the heterogeneity inherent in primary tissues and tumors and provide the means to characterize complex phenotypes, isolate rare populations, and dissect underlying mechanisms. Especially critical for cancer research is the ability to track mutations and epigenetic events that confer hallmark attributes required for aggressive growth, malignancy, and therapeutic resistance (Hanahan and Weinberg 2011). These changes impact network architecture and confer signatures that can be associated at the single-cell level with clinical features of each patient's disease (Irish et al. 2006a). Nearly all cellular features relevant for cancer research can now be measured on a per-cell basis (Table 1). A major advantage of a multidimensional, single-cell approach is that it allows determination of whether an abnormal trait in cancer, such as oncogenic signaling or a gene mutation, exists in all cells or is restricted to a cell subset (Fig. 1). As each piece of knowledge added per cell can dramatically improve the power to understand an experimental result (Krutzik et al. 2004), there has been a drive to expand the number of simultaneous per-cell measurements that can be made (Perfetto et al.2004; Bendall et al. 2011). The creation of single-cell network profiling techniques has led to important breakthroughs in blood cancer, where flow cytometry techniques are straightforward to apply (Irish et al. 2006a). There is an urgent need now to apply these tools to the challenges of early detection and analysis of solid tumor cell signaling, tumor immunity, transformation to aggressive disease, and metastasis. High-dimensional flow cytometry approaches complement rapidly developing multiplex imaging cytometry tools (Gerner et al. 2012; Gerdes et al. 2013) and single-cell genetic tools (Kalisky and Quake 2011; Wu et al. 2014). The promise of these techniques for precision medicine is great, but they also create the challenge of integrating results from multiple high-dimensional, single-cell quantitative techniques. Here we provide a primer for applying high-dimensional, single-cell flow cytometry in translational cancer research.

**Table 1** Detecting cancer hallmarks in single cells

| Cell property [a] | Example flow cytometry method (and referenced use in cancer) |
|---|---|
| Differentiation/lineage | Antibodies against c-KIT (Wozniak and Kopec-Szlezak 2004), CD34 (stem cells) (Holyoake et al. 1999; Wozniak and Kopec-Szlezak 2004; Robillard et al. 2005); antibodies against CD38 (Robillard et al. 2005) or CD20 (Robillard et al. 2005; Irish et al. 2010) and other cluster of differentiation (CD) antigens in human (Mason et al. 2002; van Dongen et al. 2012; Amir el et al. 2013) and mouse (Van Meter et al. 2007; Mayle et al. 2013) tumor and blood cancer tissue samples |
| DNA content (aneuploidy, DNA fragmentation) | PI (O'Brien and Bolton 1995), ethidium monoazide (O'Brien and Bolton 1995), or 7-actinomycin D (7-AAD) (O'Brien and Bolton 1995; Holyoake et al. 1999) staining of DNA; flow cytometry and FISH to evaluate telomere length (Baerlocher et al. 2006); γH2AX foci indicating DNA double-strand break repair (Huang et al. 2003; Bourton et al. 2012); rhodium and iridium metal intercalators (Ornatsky et al. 2008) |
| RNA content (quiescence) | Pyronin Y (Holyoake et al. 1999) staining of RNA |
| Cell cycle stage | Antibodies against cyclinD (Holyoake et al. 1999), cyclin A (Juan et al. 1998), cyclin B1 (Juan et al. 1998), cyclin E (Erlanson and Landberg 1998); phosphorylated histone H3 (M phase) (Juan et al. 1998); all cell cycle stages (Behbehani et al. 2012) |
| Proliferation | BrdU staining for newly replicated DNA (Robillard et al. 2005); antibodies against proliferating cell nuclear antigen (PCNA) (Castillo et al. 2000), antibodies against Ki67 (Holyoake et al. 1999; Castillo et al. 2000); carboxy-fluorescein diacetate succinimidyl ester (CFSE) dye (Cooperman et al. 2004) |
| Oncogene expression | Antibodies against BCL2 (Laane et al. 2005; Robillard et al. 2005; Irish et al. 2010), c-MYC (Morkve et al. 1992), RAS (Andreeff et al. 1986) |
| Mutations | Antibodies against mutant p53 (Zheng et al. 1999), H-Ras-Val12 (Carney et al. 1986) |
| Tumor suppressor activity | Antibodies against p53 protein (Zheng et al. 1999; Krutzik et al. 2004) or p21/Waf1 promoter activity driving GFP (p53R-GFP system) (Ohtani et al. 2004); antibodies against phosphorylated p53 (Krutzik et al. 2004; Irish et al. 2007) or phosphorylated Rb (Behbehani et al. 2012) |
| Apoptotic cell death | Antibodies against Caspase 3 cleavage products (Belloc et al. 2000) |
| Cell membrane changes, viability, and necrosis | AnnexinV (Belloc et al. 2000) staining for extracellular phosphatidylserine exposure, which occurs on apoptotic cells; detection of membrane permeability by PI dye exclusion (Nicoletti et al. 1991) or Alexa dye exclusion (Table 2); cisplatin exclusion (Fienberg et al. 2012) |

(continued)

**Table 1** (continued)

| Cell property [a] | Example flow cytometry method (and referenced use in cancer) |
|---|---|
| Metabolism and Redox State | Dichlorofluoresceine diacetate (DCF-DA) staining (Armstrong et al. 2002), a measure of oxidation; monobromobimane (MBrB) staining (Chow and Hedley 1995), a measure of glutathione; lipophilic fluorochrome dihexaoxacarbocyanine iodide (DiOC6(3)) (Belloc et al. 2000), a measure of mitochondrial membrane potential; mitochondria peroxy yellow 1 (MitoPY1), a fluorescent probe to quantify hydrogen peroxide levels in living cells (Dickinson and Chang 2008) |
| Tumor antigens | Antibodies against B (Timmerman et al. 2002) or T (Maecker and Levy 1989) cell receptor idiotype; tetramers against tumor antigen (e.g., tyrosinase) specific T cells (Lee et al. 1999) |
| Signaling activity | Antibodies against phosphorylated STAT and MAPK proteins (Irish et al. Irish et al. 2004; Van Meter et al. 2007; Kotecha et al. 2008), phosphorylated NF-$\kappa$B, AKT, S6, Src family kinases (SFKs), and many more (Irish et al. 2010; Bendall et al. 2011); response to drug treatment (Krutzik et al. 2008; Bodenmiller et al. 2012); Indo-1 staining for $Ca^{++}$ flux (Trentin et al. 2004); antibodies against IL-12 (Panoskaltsis et al. 2003), IFN-$\gamma$ (Lee et al. 1999) or other cytokines |

[a] Deep profiling enables >36 of such features to be measured on single cells (Bendall and Nolan 2012; Bendall et al. 2012). Adapted from (Irish et al. 2006a)

## 2 Single-Cell Quantification of Cancer Hallmarks

A vast array of cellular features can now be detected by flow cytometry (Table 1). Using mass cytometry and other high-dimensional techniques enables sets of 30 or more of these features to be measured at the single-cell level simultaneously. Each new feature measured brings the potential to better dissect the cellular heterogeneity of a tumor (Fig. 1). These features can be generally categorized as markers of cell identity, surrogate markers, and effectors. Effectors differ from surrogate markers in that they directly measure a mechanistically important part of a cellular process such as signaling (MEK phosphorylation), apoptosis (caspase 3 cleavage), or proliferation (cyclin D expression). Surrogate markers have been shown to correlate with an outcome under some circumstances but they are not thought to be effectors of that outcome. An example of a surrogate marker of cancer stem cells is CD133: CD133 does not confer stemness but rather tends to enrich for cancer stem cells. High-dimensional single-cell analysis allows simultaneous quantification of many effectors of different cellular processes in all major cell types present in a sample.

In addition to measuring extracellular antigens or using live-cell permeable, nontoxic reagents, cytometry can quantify intracellular molecules and signaling activity in fixed and permeabilized cells, allowing targets in the cytoplasm and nucleus to be detected. Examples of intracellular targets include proteins with roles in metabolic potential (Armstrong et al. 2002; Chow and Hedley 1995 and Belloc

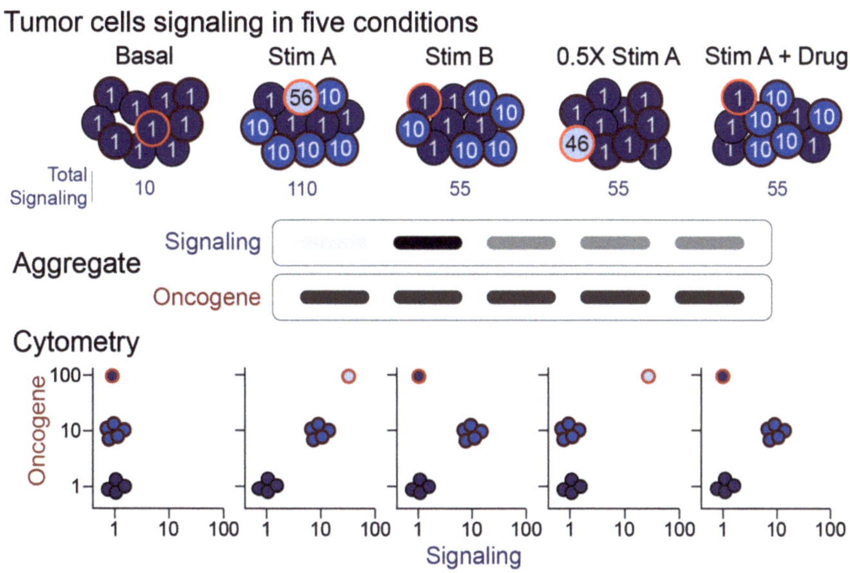


**Fig. 1** Multidimensional single-cell analysis pinpoints tumor cell signaling. In this example of 10 representative tumor cells analyzed under five stimulation conditions, oncogene expression marks three distinct populations of cells with contrasting signaling responses. In the *top row*, the number in each cell indicates the level of signaling in that cell under each condition. These values lead to the results shown as "Signaling." An aggregate analysis might mistakenly be interpreted to suggest that three of the conditions (Stim B, 0.5 × Stim A, and Stim A + Drug) elicited the same signaling responses. However, the single-cell view reveals key subset-specific signaling differences. For example, the signal from Stim B is not half as effective as Stim A. Stim B is completely effective at stimulating one subset and ineffective at stimulating another. The oncogene-high cells are hypersensitive to Stim A and nonresponsive to Stim B. Similarly, the partial effect of the Drug is due to complete inhibition of one subset and no inhibition of another. Adapted from (Krutzik et al. 2004)

et al. 2000), phosphorylation-induced signal transduction (Irish et al. 2004), and cytokine secretion (Panoskaltsis et al. 2003 and Lee et al. 1999).

As the technology to measure signaling has developed, it has aided in the development of computational modeling of biological networks in cancerous and healthy cells (Sachs et al. 2005). With the ability to quantitatively measure large sets of features simultaneously, this could lead to the systematic identification of clinically relevant signaling targets in a precision medicine setting where therapy is matched to the exact changes observed in the patient's cells. A single-cell view is critical to this, as drug responses in cell subsets are obscured when populations are analyzed in aggregate (Fig. 1).

Although a number of techniques can be used to measure certain features of cells, pragmatic concerns direct choice in many experiments. The detection techniques available to measure these features vary greatly in the amount of crosstalk that will be observed when measuring these features in combinations. A central challenge going forward is to quantitatively measure large sets of features in ways so that

**Table 2** Exclusion viability test using Alexa 700 succinimidyl ester (Ax700-SE)

| Step | Details |
|---|---|
| Ax700-SE 50,000X stock | Dissolve 1 mg Ax700-SE in 0.5 mL dimethyl sulfoxide (DMSO) to achieve a 50,000X long-term frozen stock of 2,000 μg/mL. Store frozen and protected from water |
| Ax700-SE 500X aliquots | Prepare 500X frozen stocks of 20 μg/mL Ax700-SE in DMSO. A 20 μL aliquot is sufficient to stain approximately 50 experimental samples in 200 μL |
| Ax700-SE 50X working | Dilute Ax700-SE in DMSO to prepare a 50X of 2 μg/mL. Do not store |
| Stain[a] | Add 4 μL of 50X working stock of Ax700-SE directly to cells in suspension to achieve a final concentration of 0.04 μg/mL. Stain for 10 min; titrate as needed |
| Wash and collect | Wash with 1X PBS[b] containing 1 % bovine serum albumin (BSA) or other carrier protein. Pellet cells by centrifugation and continue with other staining steps or collect |

[a] Typically, live cells are stained prior to stimulation and no apparent impact on biology is observed. For a mass cytometry version using cisplatin, refer to (Fienberg et al. 2012)
[b] Sterile filtered phosphate buffered saline (PBS) without calcium or magnesium is recommended

crosstalk between the measured channels is minimized. For example, loss of cell membrane integrity—a common surrogate for cell viability (Table 1)—should be routinely included and can be measured in many different ways that have different impacts on experiment design. In traditional flow cytometry, exclusion of fluorescent molecules like propidium iodide (PI) (Nicoletti et al. 1991), 7-aminoactinomycin D (7-AAD) (Schmid et al. 1992), and Alexa dye succinimidyl esters (SE) (Krutzik and Nolan 2006) is commonly used to detect cells lacking an intact membrane. However, PI has very broad excitation and emission spectra that greatly limit the use of additional fluorochromes detected at >550 nm.

As an alternative to PI or 7-AAD, Alexa 700 SE (Ax700-SE) can be used as a viability test (Box 1) in a manner analogous to the fluorescent cell barcoding protocol previously described (Krutzik and Nolan 2006). The Alexa dyes can be used to minimize crosstalk from the viability detection channel into other channels or to allow staining for other targets of interest on specific channels occupied by PI or 7-AAD. Sequential use of spectrally distinct Alexa dyes can be used to track changes in viability over time. In mass cytometry, a rhodium or iridium nucleic acid intercalator (Ornatsky et al. 2008) or cisplatin (Fienberg et al. 2012) can be used in a similar manner to detect cells lacking an intact plasma membrane. Detection of dead cells is especially critical when working with necrotic tumor tissue and samples from patients undergoing therapy. While centrifugation at $\sim 180 \times g$ is typical for live cells, centrifugation at $\sim 830 \times g$ is recommended to effectively pellet dead and fixed cells.

It is often useful to measure cellular features that maintain or oppose tumor growth, such as proliferation, apoptosis, and cell cycle status (Table 1). Detection of bromo-deoxyuridine (BrdU) incorporation into newly replicated DNA (Robillard 2005) and Ki67 (Holyoake et al. 1999 and Castillo et al. 2000), a protein strictly

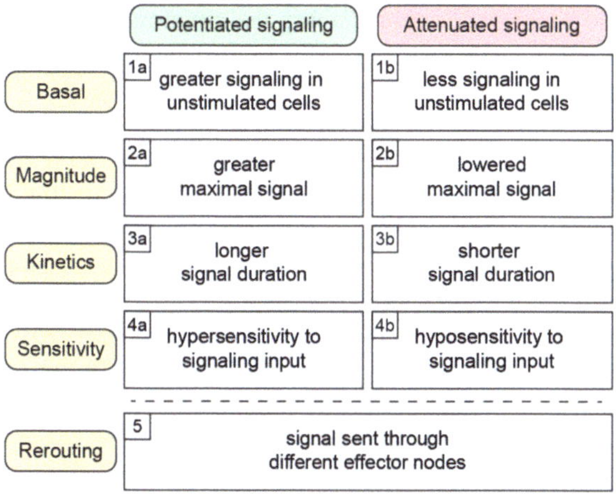


**Fig. 2** Abnormal signaling in cancer cell networks. Gains and losses of signaling drive oncogenesis and tumor progression. This figure classifies commonly observed signaling alterations according to direction (potentiated or attenuated) and mechanism. Basal signaling disruptions are commonly observed in cancer cells, and the signaling networks of the most negative prognostic cells typically display altered responses to environmental cues. Refer to (Irish et al. 2006a) for example cancer hallmark signaling changes conferred by gene mutations

associated with proliferation (Scholzen and Gerdes 2000), remain common indicators of proliferation. Apoptotic cell death is frequently measured by activation of cleaved caspase 3 or by analysis of cell membrane changes like phosphatidylserine exposure (Belloc et al. 2000). In addition, experimental drug studies with chemotherapeutics and ionizing radiation have shown that cell cycle status plays a major role in maintaining tumor homeostasis. Cytometry has explored the therapeutic implications of cells in various states of the cell cycle by revealing quiescent cells kept in a drug-tolerant state. These cells can be identified by pyronin Y staining of RNA or by the abundance of cyclins that regulate cell cycle status (Holyoake et al. 1999; Juan et al. 1998 and Erlanson and Landberg 1998). To delineate cell cycle stages by mass cytometry, 5-iodo-2-deoxyuridine (IdU) is used to mark cells in S phase and G0/G1 cells are detected using antibodies against retinoblastoma protein (Rb) phosphorylated at serines 807 and 811 (Behbehani et al. 2012).

## 3 Dissecting Abnormal Signaling Networks

Genetic and epigenetic alterations in cancer cells lead to sustained changes in basal signaling and signaling responses (Fig. 2). The vast majority of driver mutations in cancer effect profound changes in cell signaling networks (Irish et al. 2006a). These observations indicate that differential activation of signaling pathways plays a critical role in determining a cell's chance for survival or death. Epigenetic changes are also a potent force in shaping the structure of signaling networks in healthy development and cancer. Gain or loss of intercellular signaling interactions, activation of receptors whose signaling controls cell identity, and drug treatments can all trigger sustainable patterns of signaling that persist through cell division or isolation of those cells in culture. Epigenetic reprogramming of signaling networks is a primary mechanism of patterning in healthy development. B and T lymphocytes are an exception in that genetic changes are a mechanism driving healthy development and differentiation. As tools to sequence DNA and RNA continue to improve in speed, read depth, and single-cell precision (Marcy et al. 2007; Dalerba et al. 2011; Powell et al. 2012; Wu et al. 2014), genomic and proteomic tools for studying signaling network activity, transcription factor binding, and DNA methylation typically require tens of millions of cells for one test and are restricted to aggregate analysis (Fig. 1).

High-dimensional flow cytometry addresses this critical technology gap by quantifying single-cell epigenetic changes encoded by altered signaling mechanisms that transform cell function and fate (Fig. 2). Abnormal signaling in cancer can be viewed as changes in the function of signaling nodes within a network (Irish et al. 2006a). These changes are encoded by mechanisms such as constitutive basal activation of an oncogenic kinase (Fig. 2, 1a), loss of a tumor suppressor phosphatase (Fig. 2, 1b), or hypersensitivity to growth factor stimulation (Fig. 2, 4a). The signaling event can be either potentiated (strengthened) or attenuated (weakened), and these changes can have dramatic impacts on the overall function of the signaling network and the cell. Example signaling alterations in cancer that represent these mechanisms are highlighted in the following sections of this chapter.

To understand changes in regulation of signaling it is important to determine how signaling responses differ in cancer cells. A starting point to consider before analyzing a cell's entire signaling network is to identify signaling inputs that individually activate signaling nodes. In this method, cells treated with a stimulus typically serve as positive controls for signaling activity, whereas cells in a basal state function as negative controls. For constitutively active pathways, use of signaling node inhibitors may be necessary. Attention to inhibitor specificity and concentration should be considered, as the signaling response may be the result of off-target effects in a signaling network (Bodenmiller et al. 2012). With this methodology, it is possible to reveal clinically relevant signaling profiles by comparing signaling networks among patients with different clinical outcomes (Fig. 3).

There are two main phases in the generation of a validated signaling profile (Fig. 3): (1) the training phase, which has the goal of hypothesis generation and new discovery and (2) the testing phase, which is a focused challenge of a small number of hypotheses identified during testing. Development of a signaling profile begins with assembling a list of measurable features and deciding how to organize the staining panels to maximize the information gained while minimizing issues like channel crosstalk. Features are then selected according to the biosignature hypothesis, refined for clinical relevance, and tested in a new set of samples (Fig. 3).

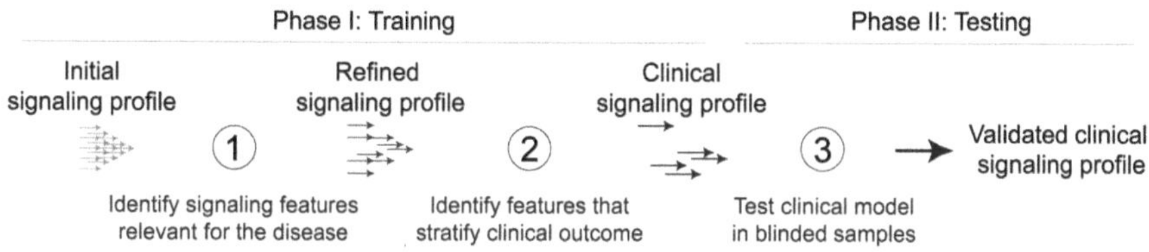


**Fig. 3** Discovery and validation of a clinical signaling profile. During the training phase, many hypotheses are tested as signaling is assessed at many nodes under a large number of conditions (basal, various signaling activators, doses, time points, drugs, and combinations). The signaling profile is then refined by determining which features differed in the experimental group (cancer) relative to controls (healthy). This feature selection step is based on the biosignature hypothesis (Irish et al. 2004), which proposes that features that vary as much in the control group as they do in the experimental group are not likely to productively contribute to unsupervised stratification because they are not specific to the experimental group. Models based on one or more features are then built, and it is determined whether they stratify an additional feature of interest that was not used to build the model, such as clinical outcome. This clinical signaling profile is then tested in a new set of samples comparable to the first and balanced for potential confounders. Ideally the test is performed by a new investigator or a computer algorithm that is blinded to the outcomes

# 4 Single-Cell Detection, Diagnosis, and Prognosis

The ability to measure multiple biomarkers per cell is particularly valuable in the study of genetically unstable tumors where new cell subsets continue to arise over time. Furthermore, cancer cells may resemble nonmalignant tumor infiltrating cells of the same lineage (Fig. 4), and detection of multiple features per cell can help clarify each cell's identity. In this example from B cell follicular lymphoma, expression of CD20, the oncogene BCL2, and BCR light chain isotype ($\kappa$ or $\lambda$) were all used to distinguish tumor B cells from nontumor host B cells. Normally B cells exhibit a mixture of $\kappa$ and $\lambda$ light chains, but in lymphoma it is common for >95 % of B cells to be a clonal expansion of a cancer cell with just one isotype. In a simple four-color panel it is possible to detect three identity markers and one phospho-protein signaling event (Fig. 4). Here, greater than normal ERK, BTK, SYK, and p38 signaling responses were identified specifically in the tumor B cells. Along with a greater magnitude of signaling potential, tumor cells sustained signaling for a significantly longer period (Irish et al. 2006b). This and other studies of BCR signaling in cancer have highlighted BCR signaling as a target for therapeutic discovery (Rickert 2013). Recently, targeting BTK has shown great promise in B cell malignancies (Byrd et al. 2013; Wang et al. 2013).

A key advantage of mass cytometry is that many surface and signaling markers can be simultaneously detected. In the fluorescence example (Fig. 4), different individual signaling readouts were repeated paired with the same three cell identity markers across four redundant staining panels in order to measure four phosphoproteins. A critical problem with this approach is that one cannot compare signaling versus signaling in the same cell—the comparison must be made at the

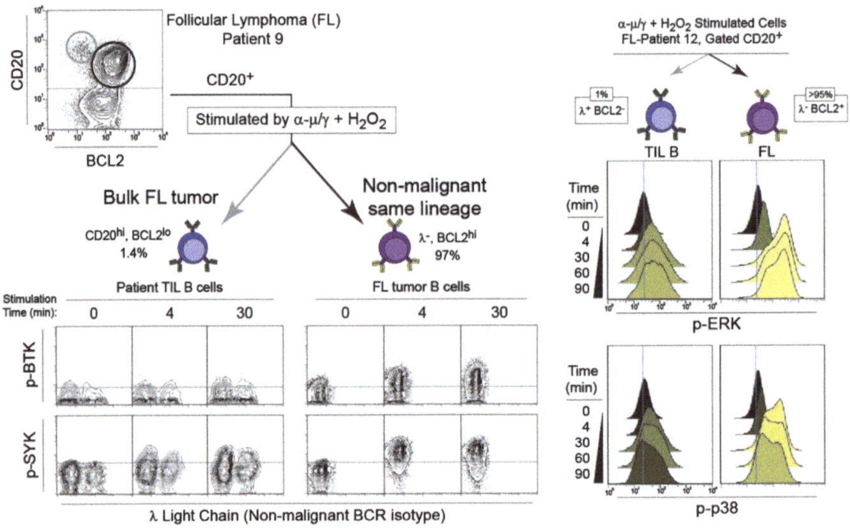


**Fig. 4** Identifying contrasting signaling in cancer and nonmalignant cells of the same lineage within a tumor. In this example, nonmalignant tumor infiltrating lymphocyte (TIL) B cells are detected within follicular lymphoma B cell tumors from two patients. On the *left*, nontumor cells were identified by the expression of the "wrong light chain"—a B cell receptor immunoglobulin light chain of a different isotype from the clonal tumor—combined with high CD20 expression and a lack of BCL2 expression. Here we can see that these cells have a distinct SYK and BTK signaling profile that contrasts with the bulk tumor. The histogram overlays on the *right* show potentiated magnitude and kinetics of ERK and p38 phosphorylation in lymphoma B cells (*right side*, BCL2+) versus TIL B cells (*left side*, identified as $\lambda+$ nontumor light chain and BCL2−)

population level. With mass cytometry, 20 markers of identity can be paired with 14 phospho-proteins in a 34-dimensional panel. This removes redundant staining panels, conserves sample, and creates higher quality data. In cases where altered signaling distinguishes cancer cells from healthy cells (Figs. 4 and 5), mass cytometry may make it possible to quickly and accurately diagnose patients based on a flow cytometry signaling profile.

Juvenile myelomonocytic leukemia (JMML) has historically been diagnosed and confirmed with a granulocyte-macrophage colony-forming units (CFU-GM) assay [(Emanuel et al. 1991) and Fig. 5]. The disadvantage of this approach is that 3–4 weeks are required to confirm the diagnosis when the potentially curative therapy for JMML is an early allogeneic stem cell transplant. While it had become clear that RAS signaling dysregulation occurs in at least 75 % of JMML (Flotho et al. 2007), the role of STAT5 activation had not been investigated. In a study that used single-cell profiling of JMML patient blood and bone marrow samples, a small proportion of CD33$^{+}$, CD14$^{+}$, CD38$^{\mathrm{dim}}$ cells exhibited hypersensitive p-STAT5 responses in response to submaximal concentrations of GM-CSF (Kotecha et al. 2008). This diagnostic approach was recently independently validated (Hasegawa et al. 2013). Thus, phospho-flow cytometry provides a precise readout for the

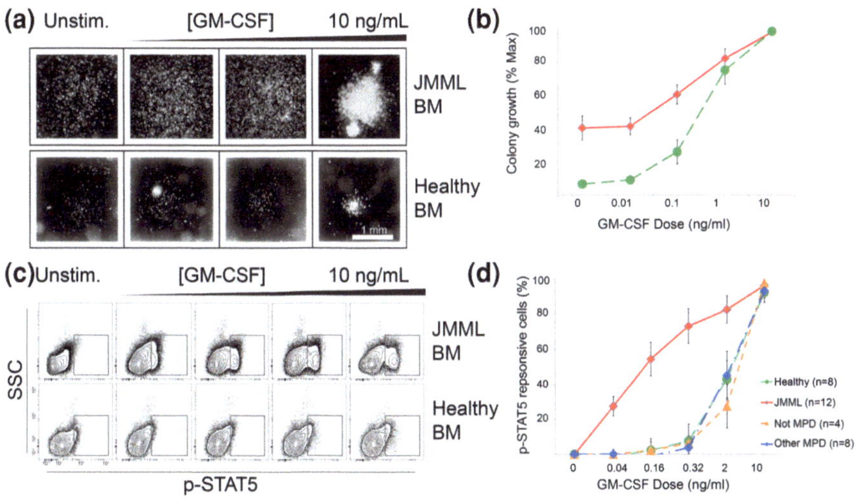


**Fig. 5** Hypersensitivity to a signaling input is diagnostic for JMML. **a** Previously, 3–4 weeks were required to confirm a suspected diagnosis of JMML with a granulocyte-macrophage colony-forming units (CFU-GM) assay. In the CFU assay, bone marrow cells from healthy donors (*green curve*) and patients have different responses to GM-CSF. **b** Plot of colony growth versus GM-CSF dose in healthy volunteers (*green*) and patients (*red*). **c** By flow cytometry, a hypersensitive population of JMML cells is detected in cancerous bone marrow compared to the normal control. **d** A dose-dependent increase in hypersensitive activity of p-STAT5 uniquely distinguished JMML from other myeloproliferative disorders as well as healthy patients. Adapted from (Kotecha et al. 2008)

aberrant signaling in JMML that distinguishes JMML from both healthy subjects and from patients with other myeloproliferative disorders. Analysis of cell subpopulations associated with disease opens opportunities for quick detection of MRD and has potential to assess therapeutic resistance (Kotecha 2008). The application to MRD is especially important in the clinical setting of cancer chemotherapy, and a vital need exists for flow cytometry tools that track and automatically identify MRD using surface markers or signaling events (Amir el et al. 2013).

## 5 Predicting Therapy Response and Tracking Evolution

Surface and signaling-based single-cell analysis can track the abundance of malignant cells at diagnosis and spot the emergence of drug-resistant cells over time during treatment. An example of this is the detection of a clinically significant tumor cell subset of lymphoma cells defined by altered BCR signaling (Fig. 6). Following α-BCR stimulation, several phospho-epitopes had impaired BCR signaling responses in a subset of cells termed lymphoma negative prognostic (LNP)

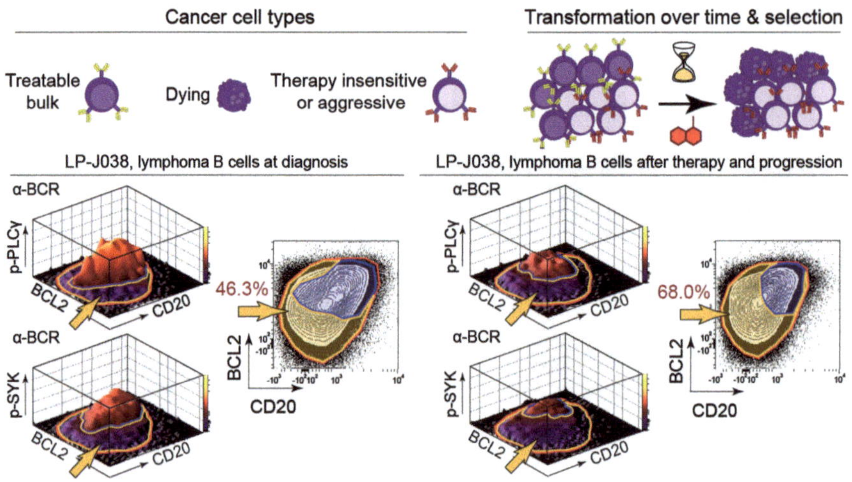


**Fig. 6** Emergence of a negative prognostic subset over time following treatment. In this example, LNP tumor cells from lymphoma patient J038 are distinguished by abnormal SYK and PLC$\gamma$ signaling and differential BCL2 and CD20 expression (*gold arrow*). At the time of diagnosis, LNP cells constituted only 46.3 % of the tumor cells. After therapy and disease progression, LNP cells increased to 68 % of the tumor. Each 1 % increase in LNP cells is associated with a 2.5 % increased risk of death in the following year ($p < 0.000005$, z-score = 4.68). Adapted from (Irish et al. 2010)

cells. The presence of BCR-insensitive LNP cells was negatively correlated with overall patient survival and LNP cells increased in abundance following treatment and disease progression (Irish et al. 2010). These results indicate that BCR-insensitive LNP cells may have a selective survival advantage compared with bulk tumor B cells (Fig. 6). The close associations between the signaling profiles and risk of death strongly suggest that these cells are therapy insensitive due to specific changes to cell signaling. Perturbing cells with an input stimulus to observe differential activation of signaling networks in cancer has repeatedly been shown to stratify survival (Irish et al. 2004, 2010).

Targeted cancer therapies have advanced rapidly as our understanding of cancer cell-specific signaling alterations has increased (Irish et al. 2006a). Genomic technologies can now identify patterns of gene expression or detect the presence of novel point mutations on a case-by-case basis. This has led to the identification of tumor subclasses and improved understanding of disease biology for appropriate therapies. For example, targeting the overexpression of HER2 with lapatinib or trastuzumab in breast cancer has benefited patients (Schnitt 2010). However, it can be difficult to target newly discovered mutations, and separating drivers from passengers can be challenging when normal, pre-, and post-treatment sample sets are not available. In contrast, the signaling events measured in phospho-flow panels are typically highly targetable, and in many cases there are drugs available that are already being used in the clinic in other settings.

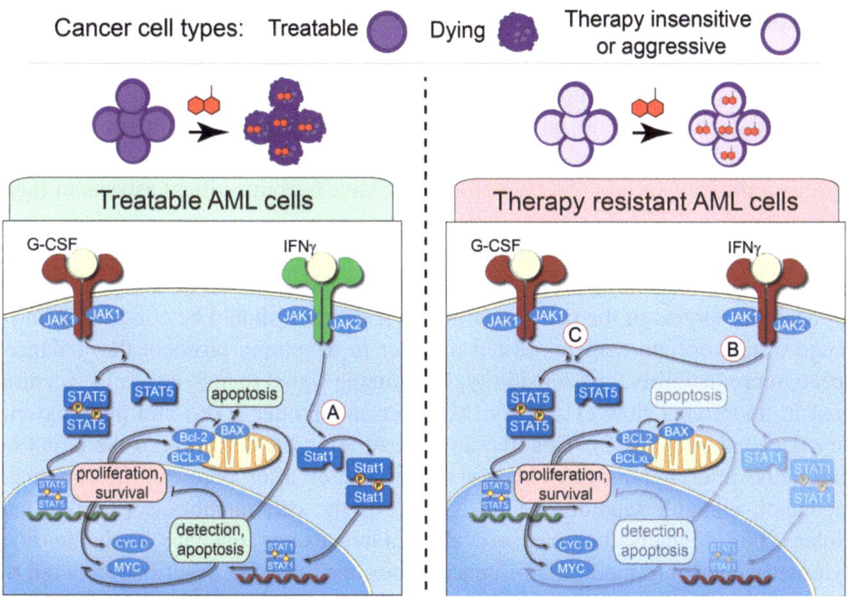


**Fig. 7** A hallmark mechanism of AML therapy resistance is rewired JAK/STAT signaling. In this example, signaling profiles of two different AML cancer cells are shown. In treatable AML cells, G-CSF signaling through JAK1 and induction of STAT5 phosphorylation mediates transcription of pro-survival and proliferation genes. Conversely, IFN$\gamma$ signaling through JAK2 results in induction of STAT1 phosphorylation that mediates cell cycle arrest and apoptosis. In the signaling network of the therapy-resistant AML cell, the response to IFN$\gamma$ has become rerouted to STAT5, which, like G-CSF, mediates transcription of pro-survival and proliferation genes. The lack of functional STAT1 activation, which activates cell cycle arrest-induced apoptosis, explains why patients with these cancer cells are often resistant to DNA-damage-induction therapy. Inhibition of JAK2/STAT5 signaling in therapy-resistant AML cells could potentially improve the outcome of patients with this resistant subset

An alternative strategy is to measure deregulation of an oncogenic pathway by measuring active kinase signaling and a cell networks signaling potential when perturbed (Fig. 7). For example, signaling alterations that predict therapy outcome are observed in acute myeloid leukemia (AML) patient samples. Increased activity of STAT5 and STAT3 activity is known to induce the expression of genes for survival and proliferation. Interferon $\gamma$ treatment activates STAT1 activity, which can oppose survival by activation of genes involved in antigen presentation to the immune system. Cells from patients who did not respond to induction chemotherapy shared a profile including a critical failure to phosphorylate STAT1 in response to interferon $\gamma$ (Fig. 7, Therapy-resistant AML cells). Instead of activating STAT1, these cells have rerouted IFN$\gamma$ signaling to phosphorylate oncogenic STAT5. These results provide a rationale for the investigation of STAT5 inhibition in therapy-resistant AML to improve the outcome of patients with this resistant subset. Thus, a key promise of the signaling profile approach is that observed cancer-specific signaling disruptions are required for cancer cell survival or aggressive behavior.

## 6 Experimental and Clinical Considerations

Cytometry can be used to quantitate multiple properties per cell that can then be correlated with biological processes or disease progression. With the ability to simultaneously measure more targets and with the ever-increasing sizes of datasets, experimental design and data processing have become critical aspects of these experiments.

A primary challenge in high-dimensional profiling of heterogeneous cells is optimization of a staining protocol that facilitates the detection of extracellular and intracellular targets of the cells. A target's localization should be considered and a range of appropriate reagents tested in order to develop a protocol that balances speed, reproducibility, and sensitivity. Optimizing signal to noise remains a central goal in fluorescent flow cytometry (Maecker and Trotter 2006) and mass cytometry (Bendall et al. 2012). This may involve titrating the detection of the target on live cells, after paraformaldehyde fixation, and/or after permeabilization (e.g., methanol, ethanol, saponin, Triton X-100) of the cell membrane (Krutzik and Nolan 2003). Panels that measure all features except one are a classic flow cytometry control termed "fluorescence minus one" (FMO), described in detail by Maecker and Trotter 2006. For mass cytometry, a comparable "mass minus one" (MMO) control is equally valuable for determining what level of signal can reliably be considered positive.

When creating multistep staining protocols for detection of extracellular and intracellular epitopes, a key advantage of small molecule dyes and the polymer metal chelators used in mass cytometry is that they are not sensitive to the common permeabilization agents. This contrasts with large protein fluorophores; fluorescence of protein fluorophores can be harmed by harsh alcohol treatments used in storage of fixed samples and during permeabilization. In mass cytometry, a multistep staining protocol is a common alternative to seeking epitope unmasking staining conditions that work well for a variety of epitopes that are localized in different cellular compartments and differentially dependent on three-dimensional conformation. In a typical signaling experiment, surface marker staining occurs after the cells have been fixed so that detection of cell identity does not alter signaling. However, since many surface marker target epitopes are no longer detectable following harsh permeabilization conditions, surface staining occurs immediately following the short fix step that stops signaling in the phospho-flow protocol (1.6 % paraformaldehyde for 5 minutes at room temperature). Thus, surface staining occurs following stimulation/fixation and prior to methanol permeabilization. For more information, see Table 1 (Krutzik et al. 2005) and Fig. 2 (Krutzik and Nolan 2003). For certain intracellular targets—especially transcription factors—permeabilization with saponin or Triton X-100 can yield superior staining. Usually a short formaldehyde fix ($\leq$10 min) does not destroy target epitopes and detection of surface markers is decreased by an acceptable $\sim$10 % of the original signal, although there are exceptions.

Antibody titration and staining optimization should follow well-established guidelines (Box 1). It is critical to titrate antibodies in the exact conditions that they will be used and to include populations of positive and negative control cells at known ratios. The stain index between positive and negative cells allows verification of the subset pattern. It is not sufficient to titrate an antibody on a uniform positive population while using unstained or isotype control stained cells as a comparison point. It is acceptable for the positive and negative cells to be in different tubes, but the advantage of staining all the cells simultaneously in multidimensional cytometry is lost. With intracellular work: less is more. Problems are typically due to over staining, which leads to nonspecific background signal [see Fig. 2 in (Krutzik and Nolan 2003)]. Antibodies that work well by immunofluorescence nearly always are suitable for fluorescent flow and mass cytometry when the same fixation, permeabilization, and staining conditions are used.

Box 1 Guidelines for titrating antibodies

(1) Titrate antibodies in house using actual experimental conditions.
(2) Mix positive and negative cells to create a signature pattern for titrations.
(3) Use well-characterized cells for titrations (not rare cells of interest).
(4) Select optimal instrument channels for titrating reagents.
(5) It may be necessary to titrate multiple clones under multiple perm conditions for intracellular epitopes that have not been widely studied.

For all types of cytometry, internal biological control populations are ideal controls. Intracellular controls transform the cellular heterogeneity that confounds aggregate approaches (Fig. 1) into a distinct advantage of single-cell approaches. Markers of stemness, such as CD34 (Woziniak and Kopec-Szlezak 2004), and lineage-restricted molecules expressed during differentiation (Mason et al. 2002) help determine the identity of tumor cells. However, developmental programs can be aberrantly activated or suppressed in both the cancer cells and the surrounding microenvironment. Because phenotypic plasticity characterizes cancer, it is especially valuable to have multiple markers that are expected to be positive and negative on each major tumor and host-cell population. A general rule is to include two positive markers and one negative marker for each major tumor and host cell type (Irish et al. 2010). Negative markers help rule out artifacts. In immune cancers, markers of clonality can be used to confirm cancer cell identity or dissect cancer cell lineage (Irish et al. 2006b; Sachen et al. ; Green et al. 2013). Cell isolation by fluorescence-activated cell sorting followed by sequencing for oncogenic mutations can confirm the id2012entity of cancer cells or be used to identify underlying driver and passenger mutations (Green et al. 2013). Ultimately, the more features detected (Table 1), the more confidence one has in the identity and biology of a given population during the discovery or training phases of a project (Fig. 3). Cytometry provides the toolkit for tracking and characterizing the ubiquitous heterogeneity of cancer.

The dysregulated intracellular signaling observed in cancer cells contrasts greatly from signaling in normal cells. Challenging of the cancer cells with perturbation reagents can reveal divergent response patterns. Even when the mechanism is not directly inferable, analysis of multiple signaling events can identify the point in a cellular system that is dysregulated. To develop a protocol that profiles signaling responses, comparison to a healthy population of cells, such as peripheral blood mononuclear cells or a tractable genetically modified cell line, often establishes a comparison point for how intracellular systems should behave.

After acquiring a large dataset, data interpretation can be a challenging hurdle in high-dimensional experiments. Traditional multiparameter techniques like flow cytometry have relied upon two-dimensional plots to visualize the data to understand correlations between the parameters. Unfortunately, as the number of parameters increase, the number of two-dimensional plots increase to create an overwhelming visualization problem. Analytical approaches developed to tackle this complexity include dimensional reduction tools such as SPADE (Bendall et al. 2011; Qiu et al. 2011) and viSNE (Amir el et al. 2013). To achieve a greater understanding of tumor proteins and signaling, these tools can be used to then computationally compare this new view of cancer across patients and tumor subtypes.

## 7 Future Perspectives

Going forward, the field must address a number of challenges in data analysis and platform integration raised by the increased power to simultaneously detect many features of single cells. Four key areas are:

(1) Data analysis, storage, and sharing with collaborative teams.
(2) Cross-platform comparisons with other systems biology techniques.
(3) Cross-scale data integration, especially between single cells and aggregates.
(4) Comparisons across time, especially in clinical studies.

Technical tools and experimental designs have far outpaced the existing computational tools. Many are working to address this need, but it is important to go beyond the basic challenge of clustering groups of cells by similar features. Tools for identification of populations within single-cell datasets have increased dramatically in sophistication and speed (Pyne et al. 2009; Qiu et al. 2011; Aghaeepour et al. 2013; Amir el et al. 2013), and now there is an urgent need for tools that model the populations and derive biological meaning from the markers used to find populations. It is critical to make sure that tools do not find populations in such a way that they are limited to a particular dataset. This is vital for reproducibility as well as for clinical application. In the end, it is critical to define

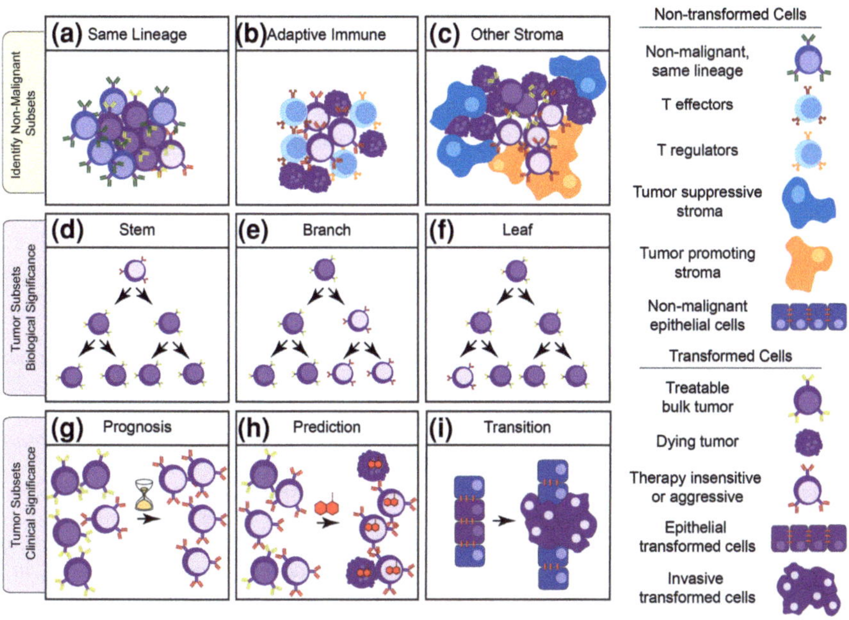


**Fig. 8** Key single-cell opportunities in cancer research. The *first row* depicts the opportunities of detecting nonmalignant cells of the same lineage as the tumor (**a**, as in Fig. 5), tumor infiltrating immune responders [**b**, as in (Myklebust et al. 2013)], and other nonmalignant stromal cells (**c**). It will be important to distinguish between abnormal signaling that promotes cancer, such as inflammation, and abnormal signaling that results from cancer, such as T cell suppression via PD-1 or generation of cancer associated fibroblasts (Barcellos-Hoff et al. 2013). In the *second row* (**d**), (**e**), and (**f**) depict contrasting biological origins of an aggressive, therapy-insensitive tumor subpopulation that can be dissected with single-cell tools. A gatekeeper mutation conferring resistance to targeted therapy might be an apomorphy that distinguishes a rare 'leaf' subset (**f**). Alternatively, a slow cell cycle phenotype might distinguish a cancer stem cell (**d**) (Reya et al. 2001). A large, heterogeneous branch (**e**) observed at the time of diagnosis might need to be treated with a combination of therapies in order to kill all populations and obtain a clinical response. The *third row* depicts clinical single-cell opportunities, such as detecting negative prognostic subpopulations (**g**, as in Fig. 7), treatment insensitive subsets (**h**), and cellular transitions like those observed when epithelial cancer cells become an invasive, metastatic population (**i**)

the difference metric in terms of the underlying biological mechanisms and to refine the model to the minimal parts for clinical testing.

How do we connect measurements made at the single-cell level with knowledge gained using other aggregate analysis tools? Single-cell techniques have essentially been developing independently of aggregate analysis tools because it is unclear as to how to connect the information gained at such different scales. Thus, approaches to span experimental platform and biological scale are sorely needed for the next generation of single-cell opportunities in cancer biology (Fig. 8).

Increasingly, single-cell tools will need to take into account changes over long periods of time—such as in the case with samples obtained over time during treatment. The concepts of before and after treatment and of subset evolution, emergence, transformation, and metastasis must be considered. What are reliable markers of stable cellular identity and how do we track "a population" of cells over time?

## References


Aghaeepour N, Finak G, Flow CAPC, Consortium D, Hoos H, Mosmann TR et al (2013) Critical assessment of automated flow cytometry data analysis techniques. Nat Methods 10:228–238

el Amir AD, Davis KL, Tadmor MD, Simonds EF, Levine JH, Bendall SC et al (2013) viSNE enables visualization of high dimensional single-cell data and reveals phenotypic heterogeneity of leukemia. Nat Biotechnol 31:545–552

Andreeff M, Slater DE, Bressler J, Furth ME (1986) Cellular RAS oncogene expression and cell cycle measured by flow cytometry in hematopoietic cell lines. Blood 67:676–681

Armstrong JS, Steinauer KK, Hornung B, Irish JM, Lecane P, Birrell GW et al (2002) Role of glutathione depletion and reactive oxygen species generation in apoptotic signaling in a human B lymphoma cell line. Cell Death Differ 9:252–263

Baerlocher GM, Vulto I, de Jong G, Lansdorp PM (2006) Flow cytometry and FISH to measure the average length of telomeres (flow FISH). Nat Protoc 1:2365–2376

Barcellos-Hoff MH, Lyden D, Wang TC (2013) The evolution of the cancer niche during multistage carcinogenesis. Nat Rev Cancer 13:511–518

Behbehani GK, Bendall SC, Clutter MR, Fantl WJ, Nolan GP (2012) Single-cell mass cytometry adapted to measurements of the cell cycle. Cytometry A 81:552–566

Belloc F, Belaud-Rotureau MA, Lavignolle V, Bascans E, Braz-Pereira E, Durrieu F et al (2000) Flow cytometry detection of caspase 3 activation in preapoptotic leukemic cells. Cytometry 40:151–160

Bendall SC, Nolan GP (2012) From single cells to deep phenotypes in cancer. Nat Biotechnol 30:639–647

Bendall SC, Nolan GP, Roederer M, Chattopadhyay PK (2012) A deep profiler's guide to cytometry. Trends Immunol 33:323–332

Bendall SC, Simonds EF, Qiu P, el Amir AD, Krutzik PO, Finck R et al (2011) Single-cell mass cytometry of differential immune and drug responses across a human hematopoietic continuum. Science 332:687–696

Bodenmiller B, Zunder ER, Finck R, Chen TJ, Savig ES, Bruggner RV et al (2012) Multiplexed mass cytometry profiling of cellular states perturbed by small-molecule regulators. Nat Biotechnol 30:858–867

Bourton EC, Plowman PN, Zahir SA, Senguloglu GU, Serrai H, Bottley G et al (2012) Multispectral imaging flow cytometry reveals distinct frequencies of gamma-H2AX foci induction in DNA double strand break repair defective human cell lines. Cytometry A 81:130–137

Byrd JC, Furman RR, Coutre SE, Flinn IW, Burger JA, Blum KA et al (2013) Targeting BTK with ibrutinib in relapsed chronic lymphocytic leukemia. N Engl J Med 369:32–42

Carney WP, Petit D, Hamer P, Der CJ, Finkel T, Cooper GM et al (1986) Monoclonal antibody specific for an activated RAS protein. Proc Nat Acad Sci USA 83:7485–7489

Castillo R, Mascarenhas J, Telford W, Chadburn A, Friedman SM, Schattner EJ (2000) Proliferative response of mantle cell lymphoma cells stimulated by CD40 ligation and IL-4. Leukemia 14:292–298

Chow S, Hedley D (1995) Flow cytometric determination of glutathione in clinical samples. Cytometry 21:68–71
Cooperman J, Neely R, Teachey DT, Grupp S, Choi JK (2004) Cell division rates of primary human precursor B cells in culture reflect in vivo rates. Stem Cells 22:1111–1120
Dalerba P, Kalisky T, Sahoo D, Rajendran PS, Rothenberg ME, Leyrat AA et al (2011) Single-cell dissection of transcriptional heterogeneity in human colon tumors. Nat Biotechnol 29:1120–1127
Dickinson BC, Chang CJ (2008) A targetable fluorescent probe for imaging hydrogen peroxide in the mitochondria of living cells. J Am Chem Soc 130:9638–9639
Emanuel PD, Bates LJ, Castleberry RP, Gualtieri RJ, Zuckerman KS (1991) Selective hypersensitivity to granulocyte-macrophage colony-stimulating factor by juvenile chronic myeloid leukemia hematopoietic progenitors. Blood 77:925–929
Erlanson M, Landberg G (1998) Flow cytometric quantification of cyclin E in human cell lines and hematopoietic malignancies. Cytometry 32:214–222
Fienberg HG, Simonds EF, Fantl WJ, Nolan GP, Bodenmiller B (2012) A platinum-based covalent viability reagent for single-cell mass cytometry. Cytometry A 81:467–475
Flotho C, Kratz C, Niemeyer CM (2007) Targeting RAS signaling pathways in juvenile myelomonocytic leukemia. Curr Drug Targets 8:715–725
Gerdes MJ, Sevinsky CJ, Sood A, Adak S, Bello MO, Bordwell A et al (2013) Highly multiplexed single-cell analysis of formalin-fixed, paraffin-embedded cancer tissue. Proc Nat Acad Sci USA 110:11982–11987
Gerner MY, Kastenmuller W, Ifrim I, Kabat J, Germain RN (2012) Histo-cytometry: a method for highly multiplex quantitative tissue imaging analysis applied to dendritic cell subset microanatomy in lymph nodes. Immunity 37:364–376
Green MR, Gentles AJ, Nair RV, Irish JM, Kihira S, Liu CL et al (2013) Hierarchy in somatic mutations arising during genomic evolution and progression of follicular lymphoma. Blood 121:1604–1611
Hanahan D, Weinberg RA (2011) Hallmarks of cancer: the next generation. Cell 144:646–674
Hasegawa D, Bugarin C, Giordan M, Bresolin S, Longoni D, Micalizzi C et al (2013) Validation of flow cytometric phospho-STAT5 as a diagnostic tool for juvenile myelomonocytic leukemia. Blood Cancer J 3:e160
Holyoake T, Jiang X, Eaves C, Eaves A (1999) Isolation of a highly quiescent subpopulation of primitive leukemic cells in chronic myeloid leukemia. Blood 94:2056–2064
Huang X, Traganos F, Darzynkiewicz Z (2003) DNA damage induced by DNA topoisomerase I- and topoisomerase II-inhibitors detected by histone H2AX phosphorylation in relation to the cell cycle phase and apoptosis. Cell Cycle 2:614–619
Irish J, Hovland R, Krutzik P, Perez O, Bruserud O, Gjertsen B et al (2004) Single cell profiling of potentiated phospho-protein networks in cancer cells. Cell 118:217–228
Irish J, Kotecha N, Nolan G (2006a) Innovation–mapping normal and cancer cell signalling networks: towards single-cell proteomics. Nat Rev Cancer 6:146–155
Irish JM, Czerwinski DK, Nolan GP, Levy R (2006b) Altered B-cell receptor signaling kinetics distinguish human follicular lymphoma. B cells from tumor-infiltrating nonmalignant B cells. Blood 108:3135–3142
Irish JM, Anensen N, Hovland R, Skavland J, Borresen-Dale A-L, Bruserud O et al (2007) Flt3 Y591 duplication and Bcl-2 overexpression are detected in acute myeloid leukemia cells with high levels of phosphorylated wild-type p53. Blood 109:2589–2596
Irish JM, Myklebust JH, Alizadeh AA, Houot R, Sharman JP, Czerwinski DK et al (2010) B-cell signaling networks reveal a negative prognostic human lymphoma cell subset that emerges during tumor progression. Proc Nat Acad Sci USA 107:12747–12754
Juan G, Traganos F, James WM, Ray JM, Roberge M, Sauve DM et al (1998) Histone H3 phosphorylation and expression of cyclins A and B1 measured in individual cells during their progression through G2 and mitosis. Cytometry 32:71–77
Kalisky T, Quake SR (2011) Single-cell genomics. Nat Methods 8:311–314

Kotecha N, Floress NJ, Irish JM, Simonds EF, Sakai DS, Archambeault S et al (2008) Single-cell profiling identifies aberrant STAT5 activation in myeloid malignancies with specific clinical and biologic correlates. Cancer Cell 14:335–343
Krutzik PO, Crane JM, Clutter MR, Nolan GP (2008) High-content single-cell drug screening with phosphospecific flow cytometry. Nat Chem Biol 4:132–142
Krutzik PO, Hale MB, Nolan GP (2005) Characterization of the murine immunological signaling network with phosphospecific flow cytometry. J Immunol 175:2366–2373
Krutzik PO, Irish JM, Nolan GP, Perez OD (2004) Analysis of protein phosphorylation and cellular signaling events by flow cytometry: techniques and clinical applications. Clin Immunol 110:206–221
Krutzik PO, Nolan GP (2006) Fluorescent cell barcoding in flow cytometry allows high-throughput drug screening and signaling profiling. Nat Methods 3:361–368
Krutzik PO, Nolan GP (2003) Intracellular phospho-protein staining techniques for flow cytometry: monitoring single cell signaling events. Cytometry A 55:61–70
Laane E, Tani E, Bjorklund E, Elmberger G, Everaus H, Skoog L et al (2005) Flow cytometric immunophenotyping including Bcl-2 detection on fine needle aspirates in the diagnosis of reactive lymphadenopathy and non-Hodgkin's lymphoma. Cytometry B Clin Cytometry 64:34–42
Lee PP, Yee C, Savage PA, Fong L, Brockstedt D, Weber JS et al (1999) Characterization of circulating T cells specific for tumor-associated antigens in melanoma patients. Nat Med 5:677–685
Maecker HT, Levy R (1989) Prevalence of antigen receptor variants in human T cell lines and tumors. J Immunol 142:1395–1404
Maecker HT, Trotter J (2006) Flow cytometry controls, instrument setup, and the determination of positivity. Cytometry A 69:1037–1042
Marcy Y, Ouverney C, Bik EM, Losekann T, Ivanova N, Martin HG et al (2007) Dissecting biological "dark matter" with single-cell genetic analysis of rare and uncultivated TM7 microbes from the human mouth. Proc Nat Acad Sci USA 104:11889–11894
Mason D, Andre P, Bensussan A, Buckley C, Civin C, Clark E et al (2002) CD antigens 2002. Blood 99:3877–3880
Mayle A, Luo M, Jeong M, Goodell MA (2013) Flow cytometry analysis of murine hematopoietic stem cells. Cytometry A 83:27–37
Morkve O, Halvorsen OJ, Stangeland L, Gulsvik A, Laerum OD (1992) Quantitation of biological tumor markers (p53, c-myc, Ki-67 and DNA ploidy) by multiparameter flow cytometry in non-small-cell lung cancer. Int J Cancer 52:851–855
Myklebust JH, Irish JM, Brody J, Czerwinski DK, Houot R, Kohrt HE et al (2013) High PD-1 expression and suppressed cytokine signaling distinguish T cells infiltrating follicular lymphoma tumors from peripheral T cells. Blood 121:1367–1376
Nicoletti I, Migliorati G, Pagliacci MC, Grignani F, Riccardi C (1991) A rapid and simple method for measuring thymocyte apoptosis by propidium iodide staining and flow cytometry. J Immunol Methods 139:271–279
O'Brien MC, Bolton WE (1995) Comparison of cell viability probes compatible with fixation and permeabilization for combined surface and intracellular staining in flow cytometry. Cytometry 19:243–255
Ohtani S, Kagawa S, Tango Y, Umeoka T, Tokunaga N, Tsunemitsu Y et al (2004) Quantitative analysis of p53-targeted gene expression and visualization of p53 transcriptional activity following intratumoral administration of adenoviral p53 in vivo. Mol Cancer Ther 3:93–100
Ornatsky OI, Lou X, Nitz M, Schafer S, Sheldrick WS, Baranov VI et al (2008) Study of cell antigens and intracellular DNA by identification of element-containing labels and metallo-intercalators using inductively coupled plasma mass spectrometry. Anal Chem 80:2539–2547
Panoskaltsis N, Reid CD, Knight SC (2003) Quantification and cytokine production of circulating lymphoid and myeloid cells in acute myelogenous leukaemia. Leukemia 17:716–730
Perfetto SP, Chattopadhyay PK, Roederer M (2004) Seventeen-colour flow cytometry: unravelling the immune system. Nat Rev Immunol 4:648–655

Powell AA, Talasaz AH, Zhang H, Coram MA, Reddy A, Deng G et al (2012) Single cell profiling of circulating tumor cells: transcriptional heterogeneity and diversity from breast cancer cell lines. PLoS One 7:e33788
Pyne S, Hu X, Wang K, Rossin E, Lin TI, Maier LM et al (2009) Automated high-dimensional flow cytometric data analysis. Proc Nat Acad Sci USA 106:8519–8524
Qiu P, Simonds EF, Bendall SC, Gibbs KD Jr, Bruggner RV, Linderman MD et al (2011) Extracting a cellular hierarchy from high-dimensional cytometry data with SPADE. Nat Biotechnol 29:886–891
Reya T, Morrison SJ, Clarke MF, Weissman IL (2001) Stem cells, cancer, and cancer stem cells. Nature 414:105–111
Rickert RC (2013) New insights into pre-BCR and BCR signalling with relevance to B cell malignancies. Nat Rev Immunol 13:578–591
Robillard N, Pellat-Deceunynck C, Bataille R (2005) Phenotypic characterization of the human myeloma cell growth fraction. Blood 105:4845–4848
Sachen KL, Strohman MJ, Singletary J, Alizadeh AA, Kattah NH, Lossos C et al (2012) Self-antigen recognition by follicular lymphoma B-cell receptors. Blood 120:4182–4190
Sachs K, Perez O, Pe'er D, Lauffenburger DA, Nolan GP (2005) Causal protein-signaling networks derived from multiparameter single-cell data. Science 308:523–529
Schmid I, Krall WJ, Uittenbogaart CH, Braun J, Giorgi JV (1992) Dead cell discrimination with 7-amino-actinomycin D in combination with dual color immunofluorescence in single laser flow cytometry. Cytometry 13:204–208
Schnitt SJ (2010) Classification and prognosis of invasive breast cancer: from morphology to molecular taxonomy. Mod Pathol 23(Suppl 2):S60–S64
Scholzen T, Gerdes J (2000) The Ki-67 protein: from the known and the unknown. J Cell Physiol 182:311–322
Timmerman JM, Czerwinski DK, Davis TA, Hsu FJ, Benike C, Hao ZM et al (2002) Idiotype-pulsed dendritic cell vaccination for B-cell lymphoma: clinical and immune responses in 35 patients. Blood 99:1517–1526
Trentin L, Cabrelle A, Facco M, Carollo D, Miorin M, Tosoni A et al (2004) Homeostatic chemokines drive migration of malignant B cells in patients with non-Hodgkin lymphomas. Blood 104:502–508
van Dongen JJ, Lhermitte L, Bottcher S, Almeida J, van der Velden VH, Flores-Montero J et al (2012) EuroFlow antibody panels for standardized n-dimensional flow cytometric immunophenotyping of normal, reactive and malignant leukocytes. Leukemia 26:1908–1975
Van Meter MEM, Diaz-Flores E, Archard JA, Passegue E, Irish JM, Kotecha N et al (2007) K-Ras(G12D) expression induces hyperproliferation and aberrant signaling in primary hematopoietic stem/progenitor cells. Blood 109:3945–3952
Wang ML, Rule S, Martin P, Goy A, Auer R, Kahl BS et al (2013) Targeting BTK with ibrutinib in relapsed or refractory mantle-cell lymphoma. N Engl J Med 369:507–516
Wozniak J, Kopec-Szlezak J (2004) c-Kit receptor (CD117) expression on myeloblasts and white blood cell counts in acute myeloid leukemia. Cytometry B Clin Cytometry 58:9–16
Wu AR, Neff NF, Kalisky T, Dalerba P, Treutlein B, Rothenberg ME et al (2014) Quantitative assessment of single-cell RNA-sequencing methods. Nat Methods 11:41–46
Zheng A, Castren K, Saily M, Savolainen ER, Koistinen P, Vahakangas K (1999) p53 status of newly established acute myeloid leukaemia cell lines. Br J Cancer 79:407–415

# Studying the Human Immunome: The Complexity of Comprehensive Leukocyte Immunophenotyping

**Angélique Biancotto and J. Philip McCoy**

**Abstract** A comprehensive study of the cellular components of the immune system requires both deep and broad immunophenotyping of numerous cell populations in an efficient and practical manner. In this chapter, we describe the technical aspects of studying the human immunome using high-dimensional (15 color) fluorescence-based immunophenotyping. We focus on the technical aspects of polychromatic flow cytometry and the initial stages in developing a panel for comprehensive leukocyte immunophenotyping (CLIP). We also briefly discuss how this panel is being used and the challenges of encyclopedic analysis of these rich data sets.

## Contents



A. Biancotto
Center for Human Immunology, Autoimmunity, and Inflammation,
National Institutes of Health, Bethesda, MD 20892, USA

J. P. McCoy (✉)
Center for Human Immunology, Autoimmunity, and Inflammation,
National Institutes of Health, 10 Center Dr, MSC 1357 Bldg 10,
Rm 8C103D, Bethesda, MD 20892, USA
e-mail: mccoyj@nhlbi.nih.gov

Current Topics in Microbiology and Immunology (2014) 377: 23–60
DOI: 10.1007/82_2013_336


Published Online: 24 August 2013

# 1 Introduction

All the genes and proteins that constitute the immune system are collectively known as the immunome; the immunome is a vastly complex and highly regulated structure that protects against infection and preserves health. An ever-increasing number of cell types comprise the immunome, and these are being defined through increasingly complex patterns of antigen expression. Technological advances over the last decade now permit high-dimensional examination of the cellular components of the immunome, making possible in-depth and broad analysis of the immune system at the same time. Such integrative, comprehensive studies will permit a much better understanding of the dynamic relationships among the myriad of cell types within the immune system than is currently possible through limited studies of only specific cell lineages or more cursory assessments of the broader system. A meaningful understanding of the immunome must begin by a study of this system in healthy individuals and analyses of the variations that exist among individuals and within each individual over time. This can then be extended to studies involving defined perturbations, such as vaccination or administration of a commonly used therapeutic such as an antibiotic, and to studies involving patients with a well-defined disease. Together, these points of information can be used to construct a meaningful database of the immunome in health, which in turn can be used to develop a deep understanding of the networks involved in responses to stimuli or in disease. It is increasingly appreciated that these studies must be conducted in humans because mouse models fail to replicate all nuances of the human immune system (Davis 2008, 2012).

Given the hundreds of cell types that have been identified in the immune system, detecting and characterizing these in an encyclopedic manner in a given individual would require collection of a large amount of blood (or other specimen) if comprehensive analyses were to be performed using low-dimensional immunophenotyping technologies. It is readily apparent that as one moves toward higher multiplicity in the number of parameters examined, the volume of specimen required for an encyclopedic study decreases proportionally. Furthermore, examination of multitudes of markers simultaneously will facilitate discovery of new cellular populations not identified by lower dimensional immunophenotyping.

The technological advances that now permit high-dimensional immunophenotyping studies can be broadly categorized into two areas: those using isotopes of lanthanide and those using fluorophores (Bendall et al. 2012). The former approach is based on mass spectroscopy, thus avoiding the use of fluorochromes and the limitations occurring due to overlap of their emission spectra in polychromatic studies. This opens the door for very high-dimensional studies—arguably as many as 100 parameters simultaneously. The trade-offs for this ability are decreased throughput compared to conventional cytometry and inability to recover the cells identified. Slow throughput can be a crucial issue, as millions or tens of millions of cells need to be analyzed when immunophenotyping specific, and likely rare, cell subpopulations. Furthermore, to have a meaningful database, immunophenotyping need to be performed on scores, if not hundreds, of samples from different subjects, thus making throughput of the assay a serious concern.

New approaches involving measurements of fluorescence take advantage of improvements in both the fluorochromes and instrument hardware. Novel fluorescent molecules ranging from quantum dot and brilliant violet stains to new viability probes have been introduced covering a wide range of excitation and emission wavelengths. Hardware improvements include low-cost, high-performance lasers in a number of wavelengths, fiber optic transmission of emitted light, and novel detectors and configurations. Such improvements have made 15–18 color immunophenotyping practical in some laboratories (Perfetto et al. 2004). Recently, a spectral analyzing cytometer has been developed that deconvolutes the spectral overlap among 32 detectors, bringing fluorescence-based cytometry to higher dimensions (Sony 2013). Although more limited in the number of potential parameters capable of examination than mass cytometry, fluorescence-based methods offer higher throughput and the ability to recover cells of interest.

Here we describe the technical aspects of high-dimensional fluorescence-based immunophenotyping as well as the early stages of our approach for comprehensive leukocyte immunophenotyping (CLIP). Our goal in using CLIP is to provide both a broad and in-depth assessment of the peripheral immune system. We opted for the fluorescence-based approach because of the higher throughput and the potential to recover populations of cells defined by complex phenotyping for functional studies. Even though the dimensionality of fluorescence-based immunophenotyping is lower than that of time-of-flight cytometry, this approach permits the identification of tens of thousands of leukocyte populations.

## 2 Technical Issues in Polychromatic Immunophenotyping

### *2.1 Instrument Specifications*

In the past few years, tremendous advancements have been made in flow cytometers. There are commercially available flow cytometers that are capable of measuring up to 20 parameters simultaneously, including 18 distinct fluorochromes as well as forward and side scatter parameters. Factors in selecting the configuration of the instrument for high polychromatic immunophenotyping include the number and types of lasers, the number of photomultiplier tubes (PMT) per laser, and the arrangement of filters and dichroic mirrors. In general, high polychromatic flow cytometry is greatly facilitated by using numerous lasers with their beams spatially separated with detectors somewhat evenly disbursed among the lasers to permit multiple fluorochromes to be excited from each laser. We utilize an instrument equipped with five lasers (355 nm, 406 nm, 488 nm, 532 nm, and 639 nm wavelengths) with 22 PMT detectors. Filters were selected to minimize fluorescence spillover from one detector to another. A full list of PMTs along with the appropriate dichroic and bandpass filters that were chosen for our CLIP panel can be found in Table 1.

Prior to evaluating the reagents and panels to be used, it is important to examined the performance of the cytometer through the use of a three-step protocol that includes optimization, calibration, and standardization of numerous components such as optical filters, dichroic mirrors (reflection and mirror transmission), laser power, laser delays, electronic noise, and window extensions (Perfetto et al. 2006, 2012). It is important to stress that any instrument to be used in such studies must have highly stable fluidics, be capable of high throughput, and be sufficient sensitivity to simultaneously detect numerous dim antigens.

## 2.2 Evaluation of Marker Expression Intensity

To obtain optimal results using a multicolor panel, numerous antibody clones directed against a specific antigens and multiple antibody-fluorochrome combinations must be screened to select reagents that yield optimal detection and staining intensities (Mahnke and Roederer 2007; Biancotto et al. 2011). Not all clones against a particular antigen recognize the same epitope, and thus may yield different patterns and intensities of staining. Therefore, one should begin by selecting clones that perform in the manner desired. We tested each antibody clone using two fluorochromes with relatively good quantum yields, phycoerythrin (PE) and allophycocyanin (APC) (Fig. 1). The clone yielding the best detection of antigen based on the separation of positive from negative staining was selected for further use. Based on the intensity of marker expressions in these experiments, antigens were categorized as markers of low (dim) expression, intermediate expression, or high expression. Various fluorochromes yield various amounts of emitted light, known as a quantum yield. High quantum yield fluorochromes are considered 'bright', whereas low quantum yield ones are considered 'dim'. A sound approach for designing high polychromatic immunophenotyping panels is to pair dim markers with bright fluorochromes and, conversely, to pair bright markers with dim fluorochromes (Mahnke and Roederer 2007). This helps to keep most markers in mid-scale and to avoid excessively high or low voltage from being applied to photomultiplier tubes (PMTs). This in turn helps to prevent the need for high levels of spectral compensation. We used this approach in designing the CLIP panel whenever possible, although practical necessities, such as the availability of

**Table 1** Instrument configuration

| Laser wavelength (nm) | Laser color | Laser power (mW) | Laser type | Spectral range for detection (nm) | Dichroic #1 (nm) | Dichroic #2 (nm) | Band Pass (nm) | Fluorochrome |
|---|---|---|---|---|---|---|---|---|
| 639 | red | 40 | cube | 750–810 | | 740LP | 780/60 | APC Cy7/APC H7 |
| | | | | 685–735 | 740LP | 685LP | 710/50 | AF700 |
| | | | | 630–670 | 685LP | | 660/20 | APC/AF647 |
| 532 | green | 150 | compass | 760–800 | | 740LP | 780/40 | PE-Cy7 |
| | | | | 685–735 | 740LP | 690LP | 710/50 | PE-Cy5.5 |
| | | | | 640–680 | 690LP | 640LP | 660/40 | PE-Cy5 |
| | | | | 600–620 | 640LP | 600LP | 610/20 | PE-TR |
| | | | | 562–588 | 600LP | | 575/25 | PE |
| 488 | blue | 50 | saphire LP | 685–735 | 685LP | 685LP | 710/50 | PerCP Cy5.5 |
| | | | | 505–525 | 505LP | 505LP | 515/20 | FITC/AF488 |
| 406 | violet | 100 | cube | 750–810 | | 740LP | 780/60 | QD800 |
| | | | | 670–740 | 740LP | 670LP | 705/70 | x |
| | | | | 640–680 | 670LP | 630LP | 660/40 | QD655 |
| | | | | 585–625 | 630LP | 595LP | 605/40 | QD605 |
| | | | | 570–595 | 595LP | 570LP | 585/42 | x |
| | | | | 540–580 | 570LP | 557LP | 560/40 | x |
| | | | | 505–535 | 557LP | 505LP | 515/20 | Aqua blue |
| | | | | 425–475 | | | 450/50 | V450 |
| 355 | ultraviolet | 60 | genesis | 500–575 | 505LP | 505LP | 525/50 | UV525 |
| | | | | 425–475 | | | 450/50 | UV450 |

*DPSS* Diode Pumped Solid State
22 PMT are in this instrument, but only 20 are useable thanks to a switch between 2 PMT on violet and ultraviolet laser

the reagents or the presence of multiple dim markers in a single tube, made some compromise a necessity.

## *2.3 Evaluation of Performance of Antibody-Fluorochrome Conjugates in Multicolor Combinations*

In addition to PE and APC, it is worth testing a number of other fluorochrome conjugates. This testing can reveal how effectively fluorochromes with different quantum yields permit discrimination of cell populations bearing the marker of interest and can give some indication of how much flexibility there is in assigning a specific fluorochrome for use with each antibody clone in a high polychromatic panel. For our CLIP staining (15 color tubes), once the intensity level of expression of our 14 markers was established, they were preliminarily tested in different panel combinations of 10 colors (a staining matrix) before moving to our final 15-color matrix. In these different staining mixtures, certain fluorochrome-antibody combinations were invariant, and others had different fluorochromes. Each staining tube was analyzed, and the final fluorochrome conjugates were selected for the antibodies based on discrimination of positive and negative populations, fluorescence intensity, percentages of positive cells, and ease of compensation. This allowed the identification of fluorochrome-antibody conjugates that yielded either optimal or suboptimal performance in a multicolor panel, and thus provided a strong rationale for the combinations to be included in the final panel (Fig. 1).

One of the goals in the development of the CLIP was to use commercial 'off the shelf' reagents whenever possible to both minimize cost and to make this panel reproducible in other laboratories. Only if commercial reagents were not available or reliable were custom-conjugated antibodies used. Before the final panel was assembled, numerous clones of antibodies and fluorochrome combinations of antibodies were tested in order to optimize the identification of cell populations and to prevent unnecessarily high amounts of spectral compensation. The feasibility of developing a panel of multiple 15-color tubes has been demonstrated in our laboratory, and the panel's utility in describing complex alterations of the immune system in the context of disease and vaccination has been demonstrated (Biancotto et al. 2012a, b).

## *2.4 Titration of Fluorochrome-Antibody Conjugates on Cells*

One of the important steps in selection of optimal fluorochrome-antibody conjugates is the titration of the selected fluorochrome-antibody conjugates with relevant cells. Most antibodies have a broad range of concentrations in which they bind to antigens. If the concentration of antibody is too low, stained cells might not

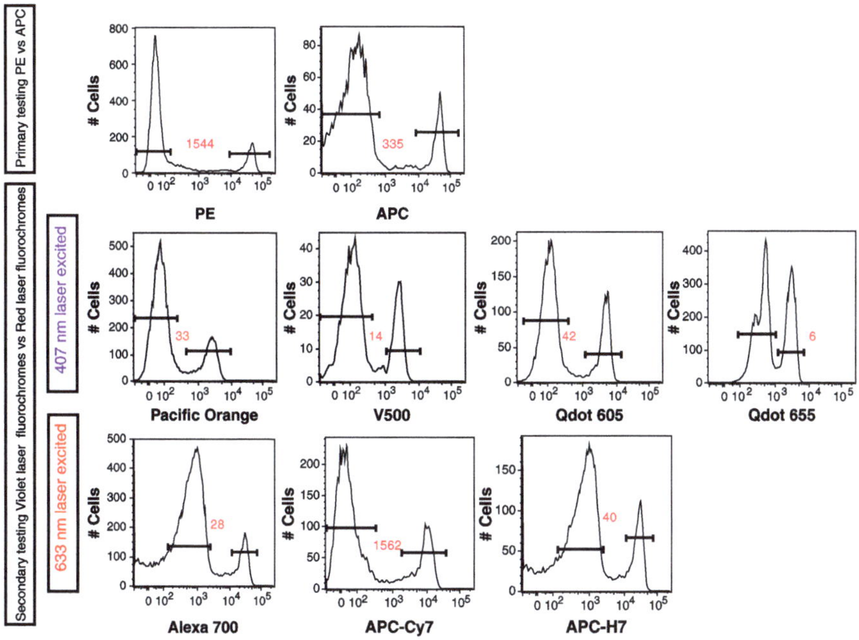


**Fig. 1** Testing of CD8 antibody clone 3B5. During the development of the CLIP panel, different fluorochrome-conjugates of CD8 antibodies were tested. Initially CD8 APC and PE were tested to measure the intensity of antigen expression. Based on the intensity of this staining, we performed secondary testing using red laser excitation (639 nm) or violet laser excitation (406 nm). Staining indices are shown in red and were calculated by dividing the MFI of the positive peak by the MFI of the negative peak

be well separated from the background or negative cells. However, if the concentration of antibody is too high, staining of the background events could increase resulting in poor separation of the positive cells from those not expressing the antigen. Therefore, the optimal concentration of antibody is that which approaches the saturation level and presents the lowest background staining, therefore giving optimal separation between positive and negative events. This will ensure a fluorescent signal that is linearly proportional to the antigen present in the sample. Thus each antibody titer was selected based on low levels of background staining, maximum separation between negative and positive populations (as measured by mean fluorescence intensity, MFI, of negative and positive populations), and for levels of positive staining close to the plateau (Fig. 2). Determining the optimal concentration of antibody to use often will decrease the cost of reagents, as titration often reveals that lower concentrations of the antibodies may be used than those recommended by the supplier. It should be stressed that titration is very much reagent and lot specific—each antibody lot that will be used in the experiment must be titered, and the use of the same clone of antibody but in a different fluorochrome conjugate requires a new titration. Titrations are usually performed as single-color assessments done independently and prior to any attempt to

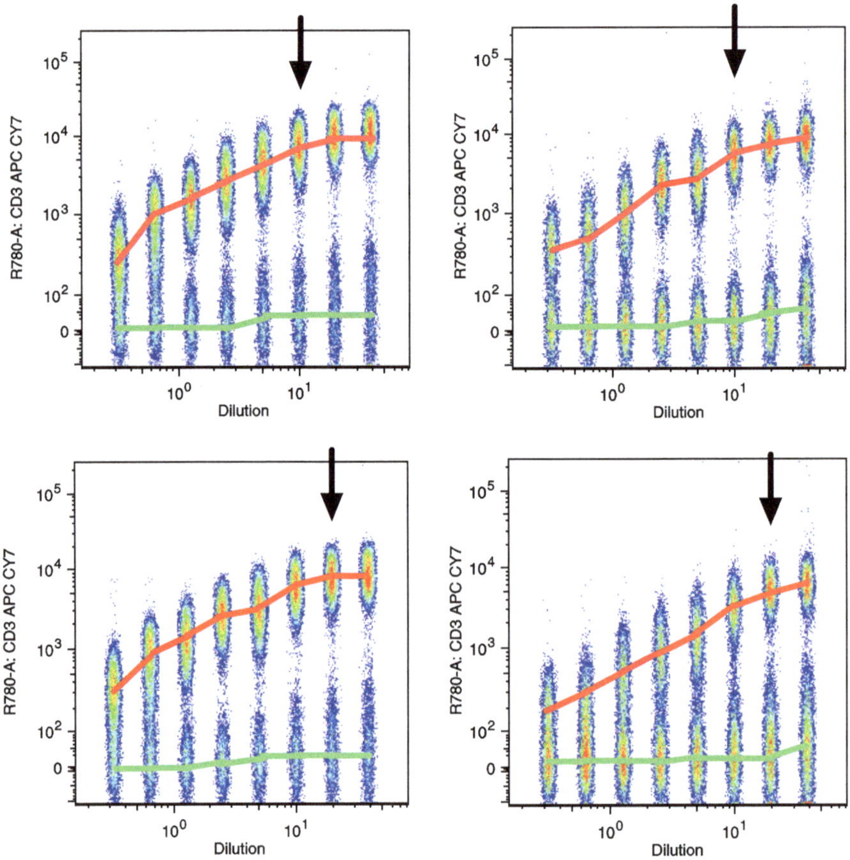


**Fig. 2** Titration of CD3 APC-Cy7. Peripheral blood mononuclear cells (PBMCs) were stained with serial dilutions of CD3 APC-Cy7 antibody. Data shown are concatenated FCS files representing serial dilution of each antibody lot used over a period of years in the CLIP panel. The cells are gated on lymphocytes (forward scatter, FSC, vs. side scatter, SSC, gates). The MFI of the positive population is marked by a *red line*. The MFI of the negative population is marked by a *green line*. Selected concentrations used in the CLIP are indicated by with an *arrow*

construct any combinations of reagents. Additionally, titration of surface markers should be performed on non-permeabilized cells whenever possible. Since this titration is not performed in combination with other antibodies or fluorochromes, titrations should be performed without compensation.

## 2.5 Fluorescence Compensation and Single-Color Controls

Multicolor flow cytometry uses several different fluorophores. At a specific excitation wavelength, each fluorophore emits light in a particular spectrum unique to that molecule, and most emission spectra exhibit a tail extending toward longer

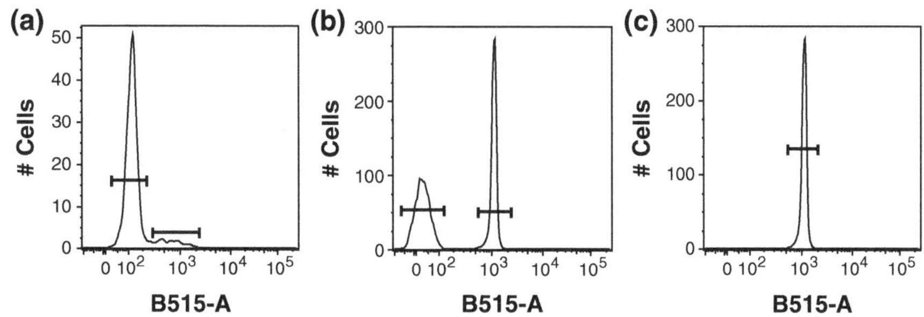


**Fig. 3** Examples of alternatives for mono-stained controls. **a** Use of cells. **b** Use of compensation beads for both the negative and positive events. **c** Use of positive compensation beads only

wavelengths, a consequence of the physics of fluorescence. In order to properly analyze multicolor flow cytometry experiments, it is necessary to assign a fluorescent signal to a particular fluorochrome and to remove any spectral spillover from all other fluorochromes. To separate fluorescence emission from the excitation light source and to resolve different fluorophores from one another, flow cytometers typically use a number of band-pass and dichroic filters, as well as spatially separated beams from excitation lasers. Even with these hardware features to separate fluorescence signals due to each fluorochrome and in spite of careful selection of the fluorochromes to be used in a panel, it is currently not possible to remove all of the spectral overlap among fluorochromes when one looks at 15 colors simultaneously. Software is used to subtract a portion of one detector's signal from another, leaving only the desired signal, a process termed fluorescence compensation. Compensation in high polychromatic flow cytometry is complex and is only practical through the use of specialized software for the purpose of determining compensation matrices. Compensation is always specific for the particular combination of specificities and fluorophores in a multicolor staining panel and, therefore, needs to be determined for each unique panel. Compensation concerns are among the foremost constraints on how many parameters can be examined simultaneously using fluorescence and have led to development of non-fluorescence-based cytometry methods such as time-of-flight cytometry. Nonetheless, when panels are properly constructed and controlled, fluorescence compensation is practical for up to 18 colors.

A matrix of the needed spectral compensation is created by running each fluorochrome-antibody conjugate individually. Although the best way to perform accurate compensation is to use the same cells that will be used in the experiments, it is difficult to accurately titer antibodies for antigens that are expressed at a low level. For this reason, one can use beads that bind to the Fc region of antibodies, and thus obtain a non-antigen binding, positive staining with high fluorescence intensity to determine compensation (Roederer 2002). For each experiment, compensation can be performed with unstained cells and compensation bead particle sets (using only the positive beads, or using positive and negative beads)

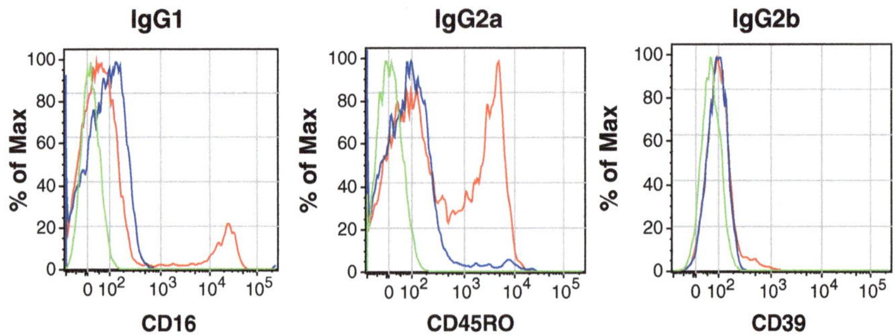


**Fig. 4** Isotype controls. Examples of overlays of unstained cells (*green line*), isotype control (*blue line*), and the antibody of interest (*red line*) for CD16 (IgG1), CD45-RO (IgG2a), and CD39 (IgG2b) antibodies. Isotype controls were used at the same concentration as the antibody of interest

(Fig. 3). In general, compensation matrices will vary little on a day-to-day basis if the same lots of properly stored reagents are used and if the cytometer in use is tightly monitored for any change in performance. For this reason, as well as the necessity of re-titrating each lot of antibodies, it is highly recommended that large lots of each reagent are purchased when performing high-dimensional immunophenotyping, thus assuring a high level of standardization across time.

## *2.6 Isotype and Fluorescence-Minus-One Control*

Unstained controls indicate the background or auto-fluorescence of cells in the sample, and isotype controls help to identify concerns that one may have about the specificity of antibody staining. The latter controls are particularly useful in intracellular staining where non-specific staining could be more problematic than in the staining of surface markers (Fig. 4). Isotype controls must be carefully matched to the specific primary antibody (species and isotype, including heavy and light chains) and to the precise fluorochrome conjugate (including the fluorophore:protein ratio if at all possible) in order to accurately determine the level of specific staining by the primary antibody. An additional type of control—fluorescence-minus-one (FMO) is used to assist in placement of quadrant gates and analysis regions. These controls contain every stain in the panel except for the one. The excluded marker is that one for which the determination of positive events may be difficult due to low intensity of staining and/or the lack of a discrete population of positive cells (Fig. 5).

## *2.7 Viability Staining*

Many studies for clinical trials are performed on cryopreserved specimens, and the viability of cells might vary for a number of reasons. Although it is difficult to

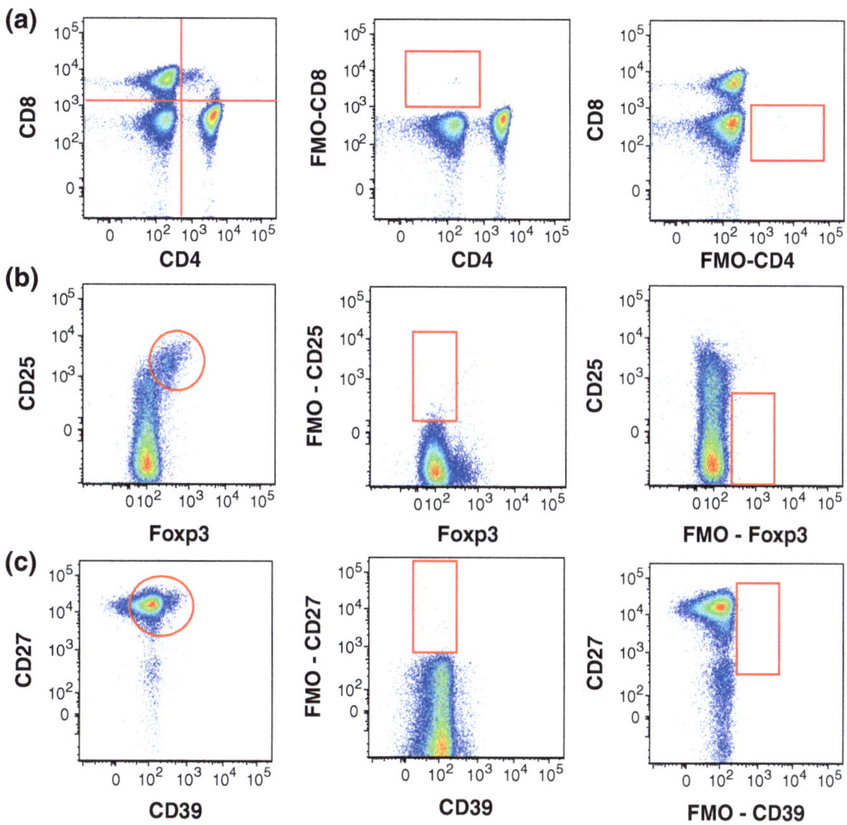


**Fig. 5** Fluorescence-minus-one control (FMO). PBMCs from the same donor were stained. Mononuclear cells were selected using forward and side light scatter parameters together with CD45 and viability staining. CD3 expression was used to identify T cells, and subsets were identified by CD4 and CD8 expression. The *boxes in red* indicate the position of the gating based on FMO. **a** Gated on T cells, *left panel* shows a bivariate dot plot of the complete staining for CD4 versus CD8; *middle panel* shows FMO control of CD8 Qd605; and *right panel* shows FMO control of CD4 V450. **b** Gated on $CD4^+$ T cells, *left panel* shows bivariate dot plot of the complete staining for regulatory T cell identification; *middle panel* shows FMO control of CD25 PE-Cy7; and *right panel* shows FMO control of Foxp3 PE. **c** Gated on T cells, *left panel* shows bivariate dot plot of the complete staining for CD27 versus CD39; *middle panel* shows FMO control of CD27 Qd655; and right panel shows FMO control of CD39 AF488. Note, that with CD25, the FMO identifies all positive cells, not only the bright cells characteristic of T regulatory cells, thus illustrating the subjectivity of place gates even with the use of FMO controls

physically remove dead cells from every sample, it is essential to be able to gate out these dead cell during analysis. This is because some antibodies bind non-specifically to dead cells present in the sample, resulting in increased fluorescence from cells not truly positive for the antigen in question. Non-specific binding may result in an overestimation of the proportion of cells that are positive for the markers of interest, and thus create false results and even falsely positive cellular populations, particularly when rare events are analyzed. For this reason, viability markers are essential and permit one to gate out dead cells. For the CLIP panel, which uses intracellular staining, a fixable dye that would not leak out of the cells after permeabilization was needed. We decided to use amine-reactive dyes as viability stains since these dyes react with free amines in the cytoplasm rather than intercalating into nucleic acids as do most other viability markers. Live cells exclude amine-reactive dyes, thanks to the integrity of the cell membrane. Dead cells take up the dye and the reaction is irreversible; the dye remains bound to the amine even when the cells are permeabilized for intracellular staining (Perfetto et al. 2010). Like all antibody-fluorochrome conjugates used in the panel, the amine-reactive dye was titered, and, even though the positive cells are gated out, this color was included in the compensation matrix (Fig. 6).

## *2.8 The Conundrum of How Many Cells are Needed for Data Acquisition*

Panels for high-dimensional immunophenotyping can be structured to provide a wide assessment of multiple lineages without in-depth analysis of each, with in-depth analysis of a particular lineage of cells, or using a combination of the two. How a particular panel is constructed will dictate the number of cells that must be acquired for statistical relevance. Acquisition counts should be predicated on collecting sufficient numbers of the rarest subset in order for a statistically meaningful analysis between samples to be conducted on that subset (generally a few hundred cells). For dendritic cell or plasmablast subsets, or even more scant populations such as endothelial progenitor cells, millions, perhaps tens of millions, of mononuclear cells must be evaluated. This presents concerns related to dimensionality. As the number of parameters increase, the depth in which any subset can be examined increases; hence, the number of cells that must be acquired increases. In turn, increasing the number of cells to be collected increases the time for data collection as well as the size of the data files, even if separate files are collected and later concatenated. Increasing collection time increases the odds of instrument fluctuations, a particular problem for instruments with low acquisition speeds. For example, collecting 10 million events at a collection rate of 20,000 events per second takes a little over 8 min, but collecting the same number of events at 500 cells per second takes roughly 5.5 h. The increased numbers of parameters acquired as well as acquiring higher number of cells needed can lead to data files of such enormity that transfer or analysis of these files on anything but the most powerful computers is agonizing slow. This issue pertains to all techniques for high-dimensional immunophenotyping and will ultimate prove to be a dimensionality barrier.

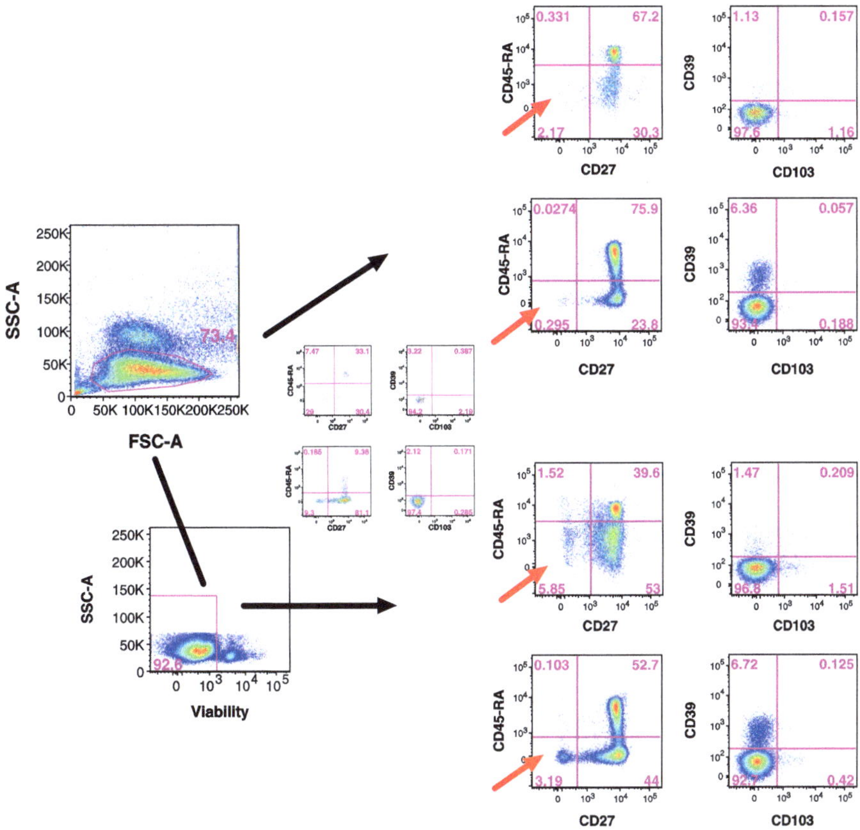


**Fig. 6** Viability staining of a sample of cryopreserved PBMCs for which we observed significant loss of cell viability upon thawing. Mononuclear cells were selected using forward and side light scatter parameters, and the expression of both CD45 and CD3 was used to identify T cells, and then CD4 and CD8 for the respective subsets (not shown). CD45-RA versus CD27 and CD39 versus CD103 expressions are shown for $CD4^+$ T cells (*upper panel*). The same analysis was performed with viability staining, in which dead cells were excluded prior detection of markers (*lower panel*). Non-specific bindings due to dead cells changed the proportion of measured populations and are indicated by a *red arrow*

## 2.9 Reproducibility and Normalization

Even when using large lots of standardized reagents, a potential concern with high-dimensional immunophenotyping, or any immunophenotyping used for clinical or translational studies, is the reproducibility of the staining over time. In many protocols, several longitudinal samples from the same patient are run at multiple time points, and it is crucial to distinguish between differences in the data that arise from biological changes from those due to assay variability. Clearly, proper instrument maintenance and assessment of its performance is a crucial element in this process; another is strict standardization in sample preparation, staining, and analysis. These issues can be addressed by strict adherence to standard operating procedures, automation of the sample processing, and possibly the use of premixed, lyophilized cocktails of the antibodies in each tube. Reproducibility of the staining panel can also be evaluated by periodically running cryopreserved aliquots from a single specimen from one donor (such as a large volume apheresis specimen that could yield hundreds of millions of cells in one specimen) over a designated period of time. These aliquots can be thawed at various intervals and stained and analyzed in a consistent manner. We have used this approach for our studies over a number of years to evaluate the reproducibility of our panel (Fig. 7). Not only does this type of analysis indicate the reproducibility of the staining, but the data collected may be used to normalize experimental data.

### 2.10 Cell Function of Immunophenotypes Defined in High Dimension

Flow cytometry is a tool of choice for the analysis of cellular phenotypes in the immune system. Deep phenotyping can and will identify novel leukocytic subsets, and whereas the function of these novels subsets might be inferred by lineage and marker expression, proof of function ultimately requires purification of these subsets for subsequent in vivo or in vitro testing. High polychromatic assays are well suited for this purpose, as cells identified in as many as 15 or more colors can be sorted by fluorescence-activated cells sorting, if a sorter is configured in the same manner as the cytometer on which the cells were identified. Novel subsets identified in higher dimension by other technologies but not recoverable, would remain of uncertain function and therefore of uncertain significance.

## 3 Comprehensive Leukocyte Immunophenotyping: One Approach

To facilitate the understanding of the immune system, a CLIP panel was developed to obtain a synoptic snapshot of the cellular phenotypes within the immune system (Biancotto et al. 2011). Historically, there have been relatively few leukocyte subsets examined in clinical studies and for which reference ranges exist (McCoy and Overton 1994). Our intent was to obtain in-depth immunophenotypic profiling of as many circulating human leukocyte populations as possible, in order to build an encyclopedic database useable by us and other groups. This was performed in concert with other concomitant measurements of the immune system such as multiplex bead array measurements of plasma cytokines concentrations, gene profiling studies, Elispot assays, and antibody titers. For immunophenotyping to

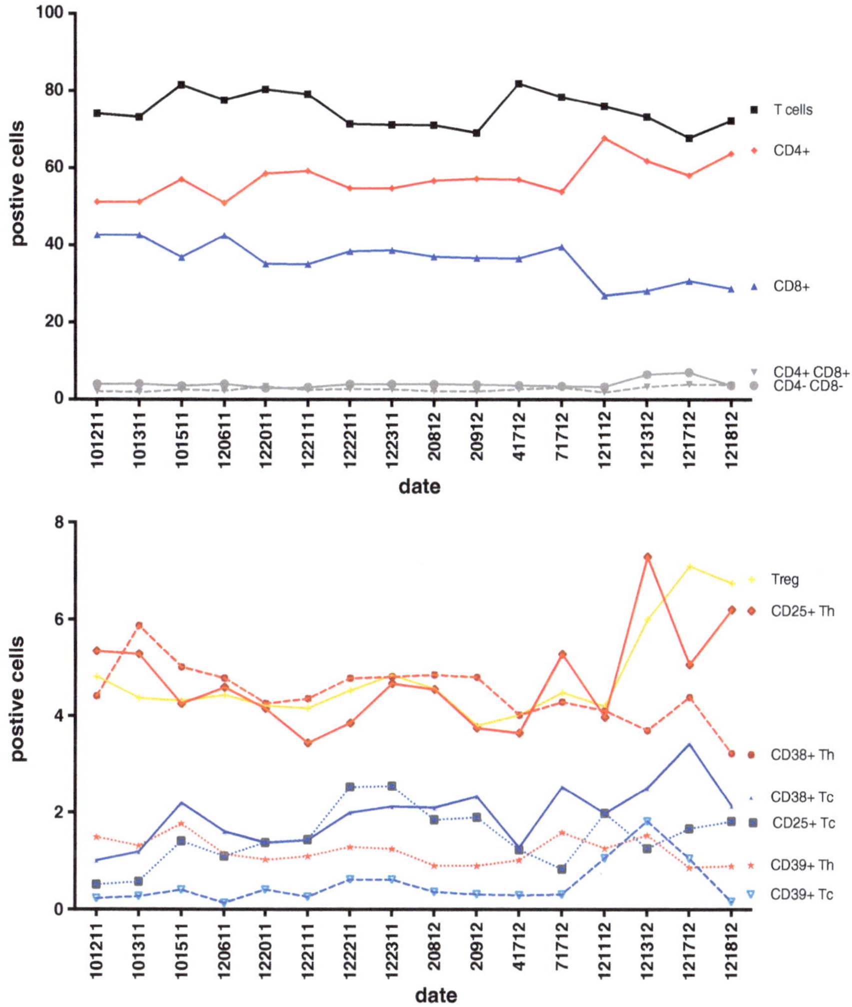


**Fig. 7** Sample normalization. A Levy-Jenning plot showing iterative analysis over a 2-year period of aliquots from a single sample (CHI-002) stained with our 15-color T regulatory cell tube. **a** T cells and T cell subpopulations $CD4^{+}CD8^{-}$, $CD4^{+}CD8^{+}$, $CD4^{-}CD8^{+}$, $CD4^{-}CD8^{-}$ are indicated. **b** Discrete markers such as activation markers on T $CD4^{+}$ cells and T $CD8^{+}$ cells and on $CD4^{+}$ regulatory T cells were identified

cover both the incredible breadth as well as fine detail of the immune system, it would be necessary to use thousands, perhaps even tens of thousands, antibodies against various cellular features. Even using the technologies for high-dimensional phenotyping that are available today, a comprehensive approach such as this would require the use of multiple individual tubes, as the number of parameters that can be measured simultaneously using any of these methods is far less than would be needed. Therefore, we reasoned that a relevant approach to construction of a CLIP panel would be to structure a series of individual tubes that each provide in-depth characterization of a specific cell type, but at the same time have a sufficient number of recurring markers among the tubes that a concatenation of the numerous phenotypes within a lineage could be assembled. For example, in our panel of 15-color tubes used to characterize T cells, there are eight redundant stains, permitting inter-tube comparisons of antigen expression among common subpopulations defined by these eight markers. Using this approach together with a literature survey of relevant markers of various leukocytes, we assembled a conceptual CLIP panel consisting of over 70 tubes of 15 parameters each. Clearly, a panel of this magnitude presents a myriad of technical concerns ranging from volume of blood to be drawn, to sample processing, to analysis of data. These will be addressed below. In spite of these concerns, our laboratory began developing tubes for this panel one at a time, in order to assess the feasibility of this approach. Here, we discuss our rationale in designing these tubes and for assembling the start of a comprehensive CLIP panel.

### *3.1 The CLIP Panel*

The academic exercise of how to immunophenotype the entire immune system is fascinating, daunting, and frustrating. Surveying the literature for all descriptive studies of the immune system using cytometry—not only for lineage and maturation markers, but also for markers of activation, subphenotypes, cytokine or chemokines production, cytokine receptors, markers of clonality such as TCR $\alpha$ and $\beta$ chains, and signaling pathways, to name a few—revealed hundreds, perhaps thousands of potential markers to be studied. Even using 15 fluorochromes in each tube, scores of tubes would be needed for such a comprehensive analysis. Using the approach described above, we began to build our CLIP panel one 15-color tube at a time. This was done with the full understanding that (1) each tube represented only one small part of a comprehensive panel, (2) that completely comprehensive analysis would not likely be achieved using the current approach, and (3) that new markers and technologies would likely evolve to replace the current efforts. Nonetheless, we felt that it was important to take steps toward comprehensive analysis to begin building a database of these findings and to learn the intricacies and caveats in such an endeavor.

To date, our panel consists of 12 15-color tubes, including three T lymphocyte tubes, three B lymphocyte tubes, two natural killer cell tubes, one dendritic cell tube, two monocytes tubes, and two neutrophil tubes. A complete list of the current tubes of the CLIP panel, along with the antibody clones and fluorochromes used can be found in Table 2. Together these tubes have the potential to identify over 49,000 subsets of leukocytes. This assumes that all combinations of markers are possible, that brightness of staining is not considered (i.e., staining is either positive or negative), and that viability, CD45, and primary lineage markers are

**Table 2** CLIP panel

| CHI CLIP panel | | | | | | | | | | |
|---|---|---|---|---|---|---|---|---|---|---|
| | | | Lymphoid | | | | | | | |
| | | | T lineage | | | B lineage | | | NK lineage | |
| **Excitation** | **Fluorochrome name** | **PMT name** | $T_1$ | $T_2$ | $T_3$ | $B_1$ | $B_2$ | $B_3$ | $NK_1$ | $NK_2$ |
| 407 nm Excitation | V450 | **V450** | **CD4** | **CD4** | **CD4** | **Lambda** | **CD80** | **CD80** | **CD56** | **CD56** |
| | | | | *RPA-T4* | *RPA-T4* | *RPA-T4* | *B39-4* | *L307.4* | *L307.4* | *B159* |
| | Aquablue | **V545** | **Viability** | **Viability** | **Viability** | **Viability** | **Viability** | **Viability** | **Viability** | **Viability** |
| | Qdot 605 | **V605** | **CD8** | **CD8** | **CD8** | **CD10** | **CD27** | **CD27** | **CD8** | **CD8** |
| | | | *3B5* | *3B5* | *3B5* | *MEM-78* | *CLB-27/1* | *CLB-27/1* | *3B5* | 3B5 |
| | Qdot 655 | **V655** | **CD27** | **CD27** | **CD27** | **CD19** | **CD19** | **CD19** | **CD30** | **NKp46** |
| | | | *CLB-27/1* | *CLB-27/1* | *CLB-27/1* | *SJ25-C1* | *SJ25-C1* | *SJ25-C1* | *BerH8* | *900* |
| | Qdot 800 | **V800** | **CD45** | **CD45** | **CD45** | **CD45** | **CD45** | **CD45** | **CD45** | **CD45** |
| | | | *HI30* | *HI30* | *HI30* | *HI30* | *HI30* | *HI30* | *HI30* | *HI30* |
| 488 nm Excitation | FITC/Alexa488 | **B515** | **CD39** | **IL-23R** | **CD69** | **CD40** | **IgA** | **IgA (intra)** | **CD57** | **CD57** |
| | | | *A1* | *218213* | *FN50* | *5C3* | *Polyclonal Goat* | *Polyclonal Goat* | *TB01* | *TB01* |
| | PcPcy5.5 | **B710** | **CD38** | **CD196** | **IL-4** | **CD138** | **CD86** | **CD86** | **CD161** | **IFNg** |
| | | | *HIT2* | *11A9* | *MP4-25D2* | *MI15* | *IT2.2* | *IT2.2* | *HP-3G10* | *4S.B3* |
| 532 nm Excitation | PE | **G560** | **Foxp3** | **CD146** | **perforin** | **Kappa** | **CD21** | **IgE (intra)** | **CD337** | **Perforin** |
| | | | *PCH101* | *142928* | *δG9* | *TB28-2* | *BL13* | *MB10-5C4* | *P30-15* | *δG9* |
| | PE-TR | **G610** | **CD45-RA** | **CD45-RO** | **CD45-RO** | **CD38** | **CD38** | **CD38** | **CD94** | **CD25** |
| | | | *2H4LDH11LDB9* | *UCHL1* | *UCHL1* | *HIT2* | *HIT2* | *HIT2* | *DX22* | *B1.49.9* |
| | PE-cy5 | **G660** | **CD103** | **CD161** | **CD40L** | **CD103** | **IgG** | **IgG (intra)** | **CD16** | **CD16** |
| | | | *LF61* | *DX12* | *TRAP1* | *LF61* | *G18-145* | *G18-145* | *3G8* | *3G8* |
| | PE-cy5.5/PE-A700 | **G710** | **HLA-DR** | **TNFa** | **TNFa** | **CD20** | **CD20** | **CD20** | **CD244** | **CD69** |
| | | | *Tu36* | *MP9-20A4* | *MP9-20A4* | *HI47* | *HI47* | *HI47* | *C1.7* | *CH/4* |
| | PE-cy7 | **G780** | **CD25** | **IL-17A** | **IFNg** | **CD5** | **CD23** | **CD23** | **CD5** | **CD5** |
| | | | *M-A251* | *BL168* | *4S.B3* | *L17F12* | *EBVCS-5* | *EBVCS-5* | *L17F12* | *L17F12* |
| 633 nm Excitation | APC/Alexa647 | **R660** | **CD127** | **IL-22** | **IL-2** | **CD22** | **IgD** | **IgD (intra)** | **CD336** | **CD127** |
| | | | | *hIL-7R-M21* | *142928* | *MQ1-17H12* | *S-HCL-1* | *IgD26* | *IgD26* | *P44-8* |
| | Alexa700/APC-cy5.5 | **R710** | **CD197** | **CD197** | **CD197** | **CD11c** | **IgM** | **IgM (intra)** | **CD158e1** | **CD158e1** |
| | | | *150503* | *150503* | *150503* | *3.9* | *CH2* | *CH2* | *DX9* | *DX9* |
| | APC-cy7/APC-H7 | **R780** | **CD3** | **CD3** | **CD3** | **CD25** | **CD10** | **CD10** | **CD3** | **CD3** |
| | | | *Sk7* | *Sk7* | *Sk7* | *M-A251* | *HI10a* | *HI10a* | *Sk7* | *Sk7* |

**Table 2** (continued)

| | | | **Myeloid** | | | | |
|---|---|---|---|---|---|---|---|
| | | | **Dendritic lineage** | **Mono lineage** | | **Neutrophil lineage** | |
| **Excitation** | **Fluorochrome name** | **PMT name** | **$DC_1$** | **$M_1$** | **$M_2$** | **$N_1$** | **$N_2$** |
| 407 Excitation | V450 | V450 | **CD123** | **HLA-DR** | **CD31** | **CD62L** | **CD14** |
| | | | *6H6* | *L243 (G46-6)* | *WM59* | *DREG-56* | *TuK4* |
| | Aquablue | V545 | **Viability** | **Viability** | **Viability** | **Viability** | **Viability** |
| | Qdot 605 | V605 | **CD27** | **CD163** | **CD163** | **CD3** | **CD8** |
| | | | *CLB-27/1* | *eBioGHI/61* | *eBioGHI/61* | *sk7* | *3B5* |
| | Qdot 655 | V655 | **CD4** | **CD14** | **CD14** | **CD14** | **CD19** |
| | | | *S3.5* | *TuK4* | *TuK4* | *TuK4* | *SJ25-C1* |
| | Qdot 800 | V800 | **CD45** | **CD45** | **CD45** | **CD16** | **CD45** |
| | | | *HI30* | *HI30* | *HI30* | *3G8* | *HI30* |
| 488 Excitation | FITC/Alexa488 | B515 | **IFNa** | **CD77** | **CD77** | **CD185** | CD123 |
| | | | *MMHA-11* | *5B5* | *GA-R2 (HIR2)* | *2G8* | |
| | PcPcy5.5 | B710 | **TNFa** | **CD192** | **CD34** | **CD203c** | **CD4** |
| | | | *MAb11* | *TG5/CCR2* | *8G12* | *NP4D6* | *S3.5* |
| 532 nm Excitation | PE | G560 | **CD40** | **CD64** | **CD64** | **CD303** | **CD56** |
| | | | *HB14* | *10.1* | *10.1* | *201A* | *NCAM16.2* |
| | PE-TR | G610 | **HLA-DR** | **CD16** | **CD16** | **CD123** | **CD16** |
| | | | *Tu36* | *3G8* | *3G8* | *6H6* | *3G8* |
| | PE-cy5 | G660 | **CD83** | **CD15** | **CD15** | **CD15** | **CD15** |
| | | | *HB15e* | *80H5* | *80H5* | *80H5* | *80H5* |
| | PE-cy5.5 /PE-A700 | G710 | **CD11c** | **CD4** | **CD4** | **HLA-DR** | **HLA-DR** |
| | | | *Bu15* | *S3.5* | *S3.5* | *Tu36* | Tu36 |
| | PE-cy7 | G780 | **lineage** | **CD13** | **CD13** | **CD13** | CD34 |

(continued)

**Table 2** (continued)

| | | | Myeloid | | | | |
|---|---|---|---|---|---|---|---|
| | | | **Dendritic lineage** | **Mono lineage** | | **Neutrophil lineage** | |
| | | | *see below* | *L138* | *L138* | *L138* | |
| 633 nm Excitation | APC/Alexa647 | R660 | **BDCA-2** | **CD36** | **CD36** | **FcεRI** | **CD11c** |
| | | | *AC144* | *CB38 (NL07)* | *CB38 (NL07)* | *BC4* | Bu15 |
| | Alexa700/APC-cy5.5 | R710 | **CD33** | **CD33** | **CD33** | **CD33** | **CD33** |
| | | | *WM-53* | *WM-53* | *WM-53* | *WM-53* | WM-53 |
| | APC-cy7/APC-H7 | R780 | **CD86** | **CD206** | **HLA-DR** | **CD10** | **CD3** |
| | | | *IT2.2* | *15-2* | *L243 (G46-6)* | *HI10a* | Sk7 |

Lineage mix =
CD3 *SK7*
CD56 *NCAM16.2*
CD20 *L27*
CD19 *SJ25C1*
CD14 *M5E2*
CD16 *3G8*

solely used for gating. Our approach was for each tube to have a specified purpose. For example, within the T lymphocyte lineage, there are specific tubes for T regulatory cells (Tregs), Th17 cells, and Th1/Th2 cells. Furthermore, as discussed above, within each lineage the tubes would have a number of recurring markers, both for quality control purposes, but also for cross-tube concatenation of more detailed phenotypes. If viable, CD45 and CD3 are used for gating in each of the T cell tubes, in theory each of the T cells tubes could yield 4,096 T cell phenotypes if one assumes all marker combinations could be expressed and that marker expression is defined only as positive or negative. Thus, although certainly not comprehensive for the immunome, the existing 12-tube CLIP panel could provide a breadth and depth of immunophenotyping not reported heretofore.

## 3.2 Applications of the CLIP Panel

The NIH Center for Human Immunology is currently using this CLIP panel for a number of translational research protocols. One early study conducted while this panel was still in development was for the characterization of Tregs in chronic lymphocytic leukemia (CLL) (Biancotto et al. 2012b). This was performed using only the one Treg tube to assess the feasibility of running and analyzing such high-dimensional data. Although several previous reports existed in the literature demonstrating elevated number of Tregs in CLL, none of these provided an in-depth characterization of Tregs found in the cancer patients compared to healthy individuals (Beyer et al. 2005; Giannopoulos et al. 2008; Deutsch et al. 2009; Jak et al. 2009; D'Arena et al. 2011). Using manual analysis, Treg populations, defined by the presence of FoxP3 in T lymphocytes, were identified not only in $CD4^{+}CD25+$ T cells, but also in $CD4^{+}CD25^{-}$ cells and CD8 lymphocytes either with or without CD25 co-expression. Furthermore, through the use of CD45RA, CD197 (CCR7), and CD27, these Treg populations were classified as naïve, effector, effector memory, or central memory cells. Within these Treg subpopulations, further subclassification was made based on the presence of CD38, CD39, CD103, CD127, and HLA-DR. A large number of these populations were present in significantly different levels in CLL patients compared to controls. Among the many differences were a population of $CD39^{+}CCR7^{+}CD4^{+}$ Tregs that were present in CLL patients but rare in healthy individuals. Because fluorescent markers were used to identify this population, it was possible to use the same staining matrix (without Foxp3 since viable Tregs were required for functional studies) to sort these cells for functional studies. This subpopulation unique to CLL patients had a lower suppressive capability than other Treg subpopulations (Fig. 8). This study revealed the power of a single tube of the CLIP panel to identify unique phenotypes within a specific cell lineage and to reveal significant differences among these detailed phenotypic subsets in health and disease and the ability of the approach to recover these phenotypes for subsequent functional studies. At the same time this study demonstrated the limitations of manual analysis of even just a

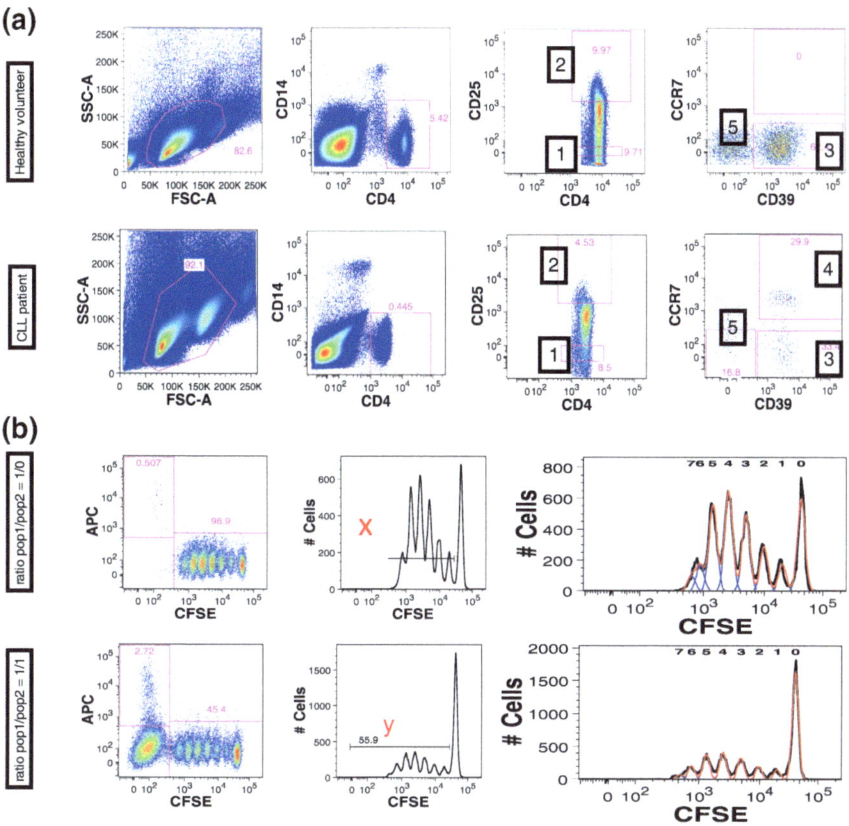


**Fig. 8** Sorting of populations identified by CLIP. **a** Based on immunophenotyping findings we stained cells in an identical manner, but without permeabilization and without FOXp3 staining, to study suppressive activity of these cells. Five populations were sorted and are identified by numbers: (1) $CD4^{+}CD25^{-}$ (effector T cells), (2) $CD4^{+}CD25^{high}$ (Treg), (3) $CD39^{+}$ Treg, (4) $CD39^{+}CCR7^{+}$ Treg, and (5) $CD39^{-}CCR7^{-}$ Treg. **b** Suppressive capacity of Treg cells on effector T cells was expressed as relative inhibition of the percentage of $CFSE^{low}$ cells according to the following formula [100 x (1− (x/y))] where $x$ is the percentage of $CFSE^{low}$ cells in the effector T cell population alone culture and $y$ is percentage of $CFSE^{low}$ cells in co-culture of Treg and effector T cells

single 15-color tube, and the near impossibility of thorough data mining without automated approaches to analysis.

Following this initial study, a second more expansive study was conducted on immune responses to an H1N1 influenza vaccine in which five of the 15-color tubes of the CLIP panel were used to assess immune responses. The tubes developed at that point included the Treg tube and tubes for Th17, Th1/2, B cells, monocytes, and dendritic cells. Cryopreserved samples from 63 patients who had received the vaccine were analyzed. Samples were collected at five time points from each patient—two pre-vaccination and three post-vaccination. As automated data analysis approaches were not yet available at the time of this study, a decision was made to select 128 subsets, out of potentially thousands, that might be the most informative for this study and perform manual analysis of these (Table 3). The guided analysis of the B cell tube alone provided 40 different populations (Fig. 9). These data were then integrated into a larger systems analysis that included gene expression data, serum cytokine levels, virus titers, and Elispot data from these same patients in order to provide a truly comprehensive assessment of immune responses to this vaccination. The results of this study have been reported elsewhere (Tsang et al. 2013). The raw flow cytometry data from this study was archived and now is being more thoroughly analyzed using a computational, rather than manual, approach (JS Tsang, personal communication).

Since the initial two studies, the CLIP panel has grown to include 12 tubes allowing study of neutrophils (Fig. 10). During this time various iterations of this panel have been used for analysis of samples from a number of clinical trials, including studies of monogenic diseases, rare undiagnosed diseases, fungal infections, and coronary artery disease (Dickler et al. 2013). A database is also being constructed using the full 12-tube CLIP panel to immunophenotype peripheral blood from healthy volunteers for use as a reference for future studies.

## *3.3 Analysis of High-Dimensional Data*

In the development of our CLIP panel, it rapidly became apparent that our ability to generate high-dimensional data exceeded our capacity to analyze the data in their entirety. Data analysis can be approached in either a 'guided' manner or in a 'discovery' mode. In the guided approach it is predetermined which leukocytes subsets will be analyzed, and gating strategies are devised based on the known expression of markers on various lineages and subpopulations. This can readily be performed manually if the number of subsets is fairly limited. In the discovery mode, a priori assumptions concerning marker expression are either ignored or limited in number, thus permitting the identification of novel, unexpected subsets. Manual analysis is of very limited use in this approach, as the numbers of potential subsets is far too great for discovery to take place in any depth.

Similar to the need for compensation software 'wizards' rather than manual setting of compensation for high polychromatic data, it is increasingly appreciated that computational approaches need to be used for the rigorous analysis of these data. If one assumes that all marker combinations are possible, and that marker expression is graded as only positive or negative, immunophenotyping with 14 antibodies could define over 8,000 leukocyte subsets. A panel of 12 tubes of 14 antibodies each could define nearly 100,000 subsets if there were no redundancies. Clearly we do not have the capacity to readily elucidate these subsets using traditional manual gating and analysis. Thus any serious attempt to enter the realm of comprehensive, high-dimensional immunophenotyping of the immunome must include automated approaches to define phenotypes in multidimensional space.

**Table 3** Flow population reported for CHI Flu study

| Population ID | Tube | Markers | Category (Page) |
|---|---|---|---|
| ID1 | T1 | CD45+CD3+ | T cells |
| ID2 | T1 | CD45+CD3+CD4+CD8− | T cells |
| ID3 | T1 | CD45+CD3+CD4+CD8+ | T cells |
| ID4 | T1 | CD45+CD3+CD4− CD8+ | T cells |
| ID5 | T1 | CD45+CD3+CD4− CD8− | T cells |
| ID6 | T1* | CD45+CD3+CD4+CD8− | T helper cells |
| ID7 | T3 | CD45+CD3+CD4+CD8−CD69+ | T helper cells |
| ID8 | T1 | CD45+CD3+CD4+CD8−CD25+ | T helper cells |
| ID9 | T1 | CD45+CD3+CD4+CD8−CD38+ | T helper cells |
| ID10 | T1 | CD45+CD3+CD4+CD8−HLA−DR+ | T helper cells |
| ID11 | T3 | CD45+CD3+CD4+CD8−CD40L+ | T helper cells |
| ID12 | T2 | CD45+CD3+CD4+CD8−CD161+ | T helper cells |
| ID13 | T2 | CD45+CD3+CD4+CD8−CD196+ | T helper cells |
| ID14 | T2 | CD45+CD3+CD4+CD8−IL17a+ | T helper cells |
| ID15 | T2 | CD45+CD3+CD4+CD8−IL21+ | T helper cells |
| ID16 | T2 | CD45+CD3+CD4+CD8−IL22+ | T helper cells |
| ID17 | T2 | CD45+CD3+CD4+CD8−IL23r+ | T helper cells |
| ID18 | T2 | CD45+CD3+CD4+CD8−CD161+ | T helper cells |
| ID19 | T2 | CD45+CD3+CD4+CD8− CD161+IL17a+ | T helper cells |
| ID20 | T2 | CD45+CD3+CD4+CD8− CD161+IL21+ | T helper cells |
| ID21 | T2 | CD45+CD3+CD4+CD8− CD161+IL22+ | T helper cells |
| ID22 | T2 | CD45+CD3+CD4+CD8−CD161− | T helper cells |
| ID23 | T2 | CD45+CD3+CD4+CD8− CD161−L17a+ | T helper cells |
| ID24 | T2 | CD45+CD3+CD4+CD8− CD161−IL21+ | T helper cells |
| ID25 | T2 | CD45+CD3+CD4+CD8− CD161−IL22+ | T helper cells |
| ID26 | T3 | CD45+CD3+CD4+CD8−IL2+ | T helper cells |
| ID27 | T3 | CD45+CD3+CD4+CD8−IFNg+ | T helper cells |
| ID28 | T3 | CD45+CD3+CD4+CD8−IL4+ | T helper cells |
| ID29.T2 | T2 | CD45+CD3+CD4+CD8−TNFa+ | T helper cells |
| ID29.T3 | T3 | CD45+CD3+CD4+CD8−TNFa+ | T helper cells |
| ID30 | T1 | CD45+CD3+CD4+CD8−CD39+ | T helper cells |
| ID31 | T1 | CD45+CD3+CD4+CD8−CD103+ | T helper cells |
| ID32 | T1 | CD45+CD3+CD4+CD8−CD127+ | T helper cells |

(continued)

**Table 3** (continued)

| Population ID | Tube | Markers | Category (Page) |
|---|---|---|---|
| ID33.T1 | T1 | CD45+CD3+CD4+CD8−CD27+ | T helper cells |
| ID33.T2 | T2 | CD45+CD3+CD4+CD8−CD27+ | T helper cells |
| ID33.T3 | T3 | CD45+CD3+CD4+CD8−CD27+ | T helper cells |
| ID34 | T1 | CD45+CD3+CD4+CD8−CD45RA+ | T helper cells |
| ID35 | T1 | CD45+CD3+CD4+CD8−CD45RA− | T helper cells |
| ID36 | T1 | CD45+CD3+CD4+CD8−CD45RA−CD27+CCR7− | T helper cells |
| ID37 | T1 | CD45+CD3+CD4+CD8−CD45RA−CD27+CCR7+ | T helper cells |
| ID38 | T1 | CD45+CD3+CD8+CD4− | T cytotoxic cells |
| ID39 | T3 | CD45+CD3+CD8+CD4−CD69+ | T cytotoxic cells |
| ID40 | T1 | CD45+CD3+CD8+CD4−CD25+ | T cytotoxic cells |
| ID41 | T1 | CD45+CD3+CD8+CD4−CD38+ | T cytotoxic cells |
| ID42 | T1 | CD45+CD3+CD8+CD4−HLA−DR+ | T cytotoxic cells |
| ID43 | T1 | CD45+CD3+CD8+CD4−CD39+ | T cytotoxic cells |
| ID44 | T1 | CD45+CD3+CD8+CD4−CD103+ | T cytotoxic cells |
| ID45 | T1 | CD45+CD3+CD8+CD4−CD127+ | T cytotoxic cells |
| ID46 | T2 | CD45+CD3+CD8+CD4−IL17a+ | T cytotoxic cells |
| ID47 | T2 | CD45+CD3+CD8+CD4−IL23r+ | T cytotoxic cells |
| ID48 | T3 | CD45+CD3+CD8+CD4−IL2+ | T cytotoxic cells |
| ID49 | T3 | CD45+CD3+CD8+CD4−IFNg+ | T cytotoxic cells |
| ID50 | T3 | CD45+CD3+CD8+CD4−Perforin+ | T cytotoxic cells |
| ID51.T2 | T2 | CD45+CD3+CD8+CD4−TNFa+ | T cytotoxic cells |
| ID51.T3 | T3 | CD45+CD3+CD8+CD4−TNFa+ | T cytotoxic cells |
| ID52.T1 | T1 | CD45+CD3+CD8+CD4−CD27+ | T cytotoxic cells |
| ID52.T2 | T2 | CD45+CD3+CD8+CD4−CD27+ | T cytotoxic cells |
| ID52.T3 | T3 | CD45+CD3+CD8+CD4−CD27+ | T cytotoxic cells |
| ID53 | T1 | CD45+CD3+CD8+CD4−CD45RA+ | T cytotoxic cells |
| ID54 | T1 | CD45+CD3+CD8+CD4−CD45RA+CD27+ | T cytotoxic cells |
| ID55 | T1 | CD45+CD3+CD8+CD4−CD45RA+CD27− | T cytotoxic cells |
| ID56 | T1 | CD45+CD3+CD8+CD4−CD45RA− | T cytotoxic cells |
| ID57 | T1 | CD45+CD3+CD8+CD4−CD45RA−CD27+CCR7+ | T cytotoxic cells |
| ID58 | T1 | CD45+CD3+CD8+CD4−CD45RA−CD27−CCR7− | T cytotoxic cells |
| ID59 | T1 | CD45+CD3+CD4+CD8−CD25highFOXp3+ | T regulatory cells |

(continued)

**Table 3** (continued)

| Population ID | Tube | Markers | Category (Page) |
|---|---|---|---|
| ID60 | T1 | CD45+CD3+CD4+CD8−CD25highFOXp3+CD38+ | T regulatory cells |
| ID61 | T1 | CD45+CD3+CD4+CD8−CD25highFOXp3+CD39+ | T regulatory cells |
| ID62 | T1 | CD45+CD3+CD4+CD8−CD25highFOXp3+CD103+ | T regulatory cells |
| ID63 | T1 | CD45+CD3+CD4+CD8−CD25highFOXp3+HLADR+ | T regulatory cells |
| ID64 | Mono/DC | CD45+CD14+ | Monocytes |
| ID65 | Mono/DC | CD45+CD14+CD40+ | Monocytes |
| ID66 | Mono/DC | CD45+CD14+CD83+ | Monocytes |
| ID67 | Mono/DC | CD45+CD14+CD86+ | Monocytes |
| ID68 | Mono/DC | CD45+CD14+HLADR+ | Monocytes |
| ID69 | Mono/DC | CD45+CD14+TNFa+ | Monocytes |
| ID70 | Mono/DC | CD45+CD3−CD1−CD20−CD16−CD56−CD14−HLADR+ | Dendritic cells |
| ID71 | Mono/DC | CD45+CD3−CD19−CD20−CD16−CD56−CD14−HLADR+TNFa+ | Dendritic cells |
| ID72 | Mono/DC | CD45+CD3−CD19−CD20−CD16−CD56−CD14−HLADR+IFNa+ | Dendritic cells |
| ID73 | Mono/DC | CD45+CD3−CD19−CD20−CD16−CD56−CD14−HLADR+CD11c+CD123− | Dendritic cells |
| ID74 | Mono/DC | CD45+CD3−CD19−CD20−CD16−CD56−CD14−HLADR+CD11c+CD123−TNFa+ | Dendritic cells |
| ID75 | Mono/DC | CD45+CD3−CD19−CD20−CD16−CD56−CD14−HLADR+CD11c+CD123−IFNa+ | Dendritic cells |
| ID76 | Mono/DC | CD45+CD3−CD19−CD20−CD16−CD56−CD14−HLADR+CD11c−CD123+ | Dendritic cells |
| ID77 | Mono/DC | CD45+CD3−CD19−CD20−CD16−CD56−CD14−HLADR+CD11c−CD123+TNFa+ | Dendritic cells |
| ID78 | Mono/DC | CD45+CD3−CD19−CD20−CD16−CD56−CD14−HLADR+CD11c−CD123+IFNa+ | Dendritic cells |
| ID79 | Mono/DC | CD45+CD3−CD19−CD20−CD16−CD56−CD14−HLADR+CD11c+CD123+ | Dendritic cells |
| ID80 | B2 | CD45+CD19+ | B cells |
| ID81 | B2 | CD45+CD19+CD20+CD80+ | B cells |
| ID82 | B2 | CD45+CD19+CD20+CD86+ | B cells |
| ID83 | B2 | CD45+CD19+CD20+IgA+ | B cells |
| ID84 | B2 | CD45+CD19+CD20+IgG+ | B cells |
| ID85 | B2 | CD45+CD19+CD20+IgM+IgD+ | B cells |
| ID86 | B2 | CD45+CD19+CD20+IgM−IgG− | B cells |
| ID87 | B2 | CD45+CD19+CD20−CD27highCD38 ++ | Plasmablast |
| ID88 | B2 | CD45+CD19+CD20−CD27highCD38 ++CD23+ | Plasmablast |
| ID89 | B2 | CD45+CD19+CD20−CD27highCD38 ++CD21+ | Plasmablast |
| ID90 | B2 | CD45+CD19+CD20+CD27−CD10+ | Transitional B cells |
| ID91 | B2 | CD45+CD19+CD20+CD27−CD10+CD38+ | Transitional B cells |
| ID92 | B2 | CD45+CD19+CD20+CD27−CD10+CD23+ | Transitional B cells |

(continued)

**Table 3** (continued)

| Population ID | Tube | Markers | Category (Page) |
|---|---|---|---|
| ID93 | B2 | CD45+CD19+CD20+CD27−CD10+CD21+ | Transitional B cells |
| ID94 | B2 | CD45+CD19+CD20+IgD−CD27+ | Mature B cells |
| ID95 | B2 | CD45+CD19+CD20+IgD−CD27+CD23+ | Mature B cells |
| ID96 | B2 | CD45+CD19+CD20+IgD−CD27+CD38+ | Mature B cells |
| ID97 | B2 | CD45+CD19+CD20+IgD−CD27+CD80+ | Mature B cells |
| ID98 | B2 | CD45+CD19+CD20+IgD−CD27+CD86+ | Mature B cells |
| ID99 | B2 | CD45+CD19+CD20+IgD−CD27+IgA+ | Mature B cells |
| ID100 | B2 | CD45+CD19+CD20+IgD−CD27+IgG+ | Mature B cells |
| ID101 | B2 | CD45+CD19+CD20+IgD+CD27+ | Mature B cells |
| ID102 | B2 | CD45+CD19+CD20+IgD+CD27+CD23+ | Mature B cells |
| ID103 | B2 | CD45+CD19+CD20+IgD+CD27+CD38+ | Mature B cells |
| ID104 | B2 | CD45+CD19+CD20+IgD+CD27+CD80+ | Mature B cells |
| ID105 | B2 | CD45+CD19+CD20+IgD+CD27+CD86+ | Mature B cells |
| ID106 | B2 | CD45+CD19+CD20+IgD+CD27− | Mature B cells |
| ID107 | B2 | CD45+CD19+CD20+IgD+CD27−CD21+ | Mature B cells |
| ID108 | B2 | CD45+CD19+CD20+IgD+CD27−CD38+ | Mature B cells |
| ID109 | B2 | CD45+CD19+CD20+IgD+CD27−CD80+ | Mature B cells |
| ID110 | B2 | CD45+CD19+CD20+IgD+CD27−CD86+ | Mature B cells |
| ID111 | B2 | CD45+CD19+CD20+IgD+CD27−IgA+ | Mature B cells |
| ID112 | B2 | CD45+CD19+CD20+IgD+CD27−IgG+ | Mature B cells |
| ID113 | B2 | CD45+CD19+CD20+IgD−CD27− | Mature B cells |
| ID114 | B2 | CD45+CD19+CD20+IgD−CD27−CD21+ | Mature B cells |
| ID115 | B2 | CD45+CD19+CD20+IgD−CD27−CD38+ | Mature B cells |
| ID116 | B2 | CD45+CD19+CD20+IgD−CD27−CD80+ | Mature B cells |
| ID117 | B2 | CD45+CD19+CD20+IgD−CD27−CD86+ | Mature B cells |
| ID118 | B2 | CD45+CD19+CD20+IgD−CD27−CD23+ | Mature B cells |
| ID119 | B2 | CD45+CD19+CD20+IgD−CD27−IgA+ | Mature B cells |
| ID120 | B2 | CD45+CD19+CD20+IgD−CD27−IgG+ | Mature B cells |

**Table 3** (continued)

| Population ID | Parent population | Subset name | Name for paper | Comments |
|---|---|---|---|---|
| ID1 | (as of viable CD45+) | T cells | % CD3+of viable CD45+cells (Total T cells) | |
| ID2 | (as of viable CD45+CD3+) quadrant | CD4+ T helper | % CD4+ of total T cells | |
| ID3 | (as of viable CD45+CD3+) | DOUBLE POSITIVE | % CD4+CD8+ of total T cells | |
| ID4 | (as of viable CD45+CD3+) quadrant | CD8+ T cytotoxic | % CD8+ of total T cells | |
| ID5 | (as of viable CD45+CD3+) | DN immature T | % CD4−CD8−of total T cells | |
| ID6 | (as of viable CD45+CD3+) gated region | CD4+ T helper | % CD4+ of total T cells | Duplicate of ID2 |
| ID7 | (as of viable CD45+CD3+CD4+CD8−) | Activated CD4+ T helper cells CD69+ | % CD69+ of CD4+ T cells | |
| ID8 | (as of viable CD45+CD3+CD4+CD8−) | Activated CD4+ T helper cells CD25+ | % CD25+ of CD4+ T cells | |
| ID9 | (as of viable CD45+CD3+CD4+CD8−) | Activated CD4+ T helper cells CD38+ | % CD38+ of CD4+ T cells | |
| ID10 | (as of viable CD45+CD3+CD4+CD8−) | Activated CD4+ T helper cells HLA−DR+ | % HLA−DR+ of CD4+ T cells | |
| ID11 | (as of viable CD45+CD3+CD4+CD8−) | Activated T helper cells CD40L+ | % CD40+ of CD4+ T cells | |
| ID12 | (as of viable CD45+CD3+CD4+CD8−) | T helper Th17 CD161+ | % CD161+ of CD4+ T cells | |
| ID13 | (as of viable CD45+CD3+CD4+CD8−) | T helper Th17 CD196+ (CCR6) | % CD196+ of CD4+ T cells | |
| ID14 | (as of viable CD45+CD3+CD4+CD8−) | T helper Th17 IL−17A+ | % IL17+ of CD4+ T cells | |
| ID15 | (as of viable CD45+CD3+CD4+CD8−) | T helper IL−21+ | % IL21+ of CD4+ T cells | |
| ID16 | (as of viable CD45+CD3+CD4+CD8−) | T helper IL−22+ | % IL22+ of CD4+ T cells | |
| ID17 | (as of viable CD45+CD3+CD4+CD8−) | T helper IL−23R+ | % IL23+ of CD4+ T cells | |
| ID18 | (as of viable CD45+CD3+CD4+CD8−) histogram | Th17 | % CD161+ of CD4+ T cells | Duplicate of ID12 |
| ID19 | (as of viable CD45+CD3+CD4+CD8− CD161+) | Th17 CD161+ IL17A+ | % IL17+ of CD161+CD4+ T cells | |
| ID20 | (as of viable CD45+CD3+CD4+CD8− CD161+) | Th17 CD161+IL21+ | % IL21+ of CD161+CD4+ T cells | |
| ID21 | (as of viable CD45+CD3+CD4+CD8− CD161+) | Th17 CD161+IL22+ | % IL22+ of CD161+CD4+ T cells | |
| ID22 | (as of viable CD45+CD3+CD4+CD8−) histogram | Th17 CD161− | % CD161− of CD4+ T cells | |
| ID23 | (as of viable CD45+CD3+CD4+CD8− CD161−) | Th17 CD161− IL17A+ | % IL17+ of CD161−CD4+ T cells | |
| ID24 | (as of viable CD45+CD3+CD4+CD8− CD161−) | Th17 CD161−IL21+ | % IL21+ of CD161−CD4+ T cells | |
| ID25 | (as of viable CD45+CD3+CD4+CD8− CD161−) | Th17 CD161− IL22+ | % IL22+ of CD161−CD4+ T cells | |
| ID26 | (as of viable CD45+CD3+CD4+CD8−) | IL−2 T helper cells | % IL2+ of CD4+ T cells | |
| ID27 | (as of viable CD45+CD3+CD4+CD8−) | Th1 | % IFNg+ of CD4+ T cells | |
| ID28 | (as of viable CD45+CD3+CD4+CD8−) | Th2 | % IL4+ of CD4+ T cells | |
| ID29.T2 | (as of viable CD45+CD3+CD4+CD8−) | TNFa T helper cells | % TNFa+ of CD4+ T cells | |

(continued)

**Table 3** (continued)

| Population ID | Parent population | Subset name | Name for paper | Comments |
|---|---|---|---|---|
| ID29.T3 | (as of viable CD45+CD3+CD4+CD8−) | TNFa T helper cells | % TNFa+ of CD4+ T cells | |
| ID30 | (as of viable CD45+CD3+CD4+CD8−) | Activated T helper cells CD39+ (Treg) | % CD39+ of CD4+ T cells | |
| ID31 | (as of viable CD45+CD3+CD4+CD8−) | Integrin aE T helper cells (Treg) | % CD103+ of CD4+ T cells | |
| ID32 | (as of viable CD45+CD3+CD4+CD8−) | IL7R on T helper cells (activated and Treg) | % CD127(IL7R)+ of CD4+ T cells | |
| ID33.T1 | (as of viable CD45+CD3+CD4+CD8−) | T helper CD27+ | % CD27+ of CD4+ T cells | |
| ID33.T2 | (as of viable CD45+CD3+CD4+CD8−) | T helper CD27+ | % CD27+ of CD4+ T cells | |
| ID33.T3 | (as of viable CD45+CD3+CD4+CD8−) | T helper CD27+ | % CD27+ of CD4+ T cells | |
| ID34 | (as of viable CD45+CD3+CD4+CD8−) | Naïve T helper cells | % CD45RA+ of CD4+ T cells (Naïve T) | |
| ID35 | (as of viable CD45+CD3+CD4+CD8−) | Total Memory T helper cells | % CD45RA− of CD4+ T cells (Total memory CD4+ T) | |
| ID36 | (as of viable CD45+CD3+CD4+CD8− CD45RA−) | Effector memory T helper cells | % CD27+ CCR7− of memory CD4+ T cells (Effector memory CD4+T) | |
| ID37 | (as of viable CD45+CD3+CD4+CD8− CD45RA−) | Central memory T helper cells | % CD27+CCR7− of memory CD4+ T cells (Central memory CD4+ T) | |
| ID38 | (as of viable CD45+CD3+) gated region | T cytotoxic | % CD8+ of total T cells | duplicate of ID4 |
| ID39 | (as of viable CD45+CD3+CD8+CD4−) | Activated T cytotoxic CD69+ | % CD69+ of CD8+ T cells | |
| ID40 | (as of viable CD45+CD3+CD8+CD4−) | Activated T cytotoxic CD25+ | % CD25+ of CD8+ T cells | |
| ID41 | (as of viable CD45+CD3+CD8+CD4−) | Activated T cytotoxic CD38+ | % CD38+ of CD8+ T cells | |
| ID42 | (as of viable CD45+CD3+CD8+CD4−) | Activated T cytotoxic HLA−DR+ | % HLA−DR+ of CD8+ T cells | |
| ID43 | (as of viable CD45+CD3+CD8+CD4−) | Activated T cytotoxic CD39+ | % CD39+ of CD8+ T cells | |
| ID44 | (as of viable CD45+CD3+CD8+CD4−) | Activated T cytotoxic CD103+ | % CD103+ of CD8+ T cells | |
| ID45 | (as of viable CD45+CD3+CD8+CD4−) | Activated T cytotoxic CD 127+ | % TNFa+ of CD8+ T cells | |
| ID46 | (as of viable CD45+CD3+CD8+CD4−) | Tc17 IL−17A+ | % IL17A+ of CD8+ T cells (Tc17) | |
| ID47 | (as of viable CD45+CD3+CD8+CD4−) | Tc17IL−23R+ | % IL23R+ of CD8+ T cells | |
| ID48 | (as of viable CD45+CD3+CD8+CD4−) | T Cytotoxic IL2+ | % IL2+ of CD8+ T cells | |
| ID49 | (as of viable CD45+CD3+CD8+CD4−) | T Cytotoxic CD8+ IFNg+ | % INFg+ of CD8+T cells | |
| ID50 | (as of viable CD45+CD3+CD8+ CD4−) | T Cytotoxic CD8+ Perforin+ | % Perforin+ of CD8+ T cells | |
| ID51.T2 | (as of viable CD45+CD3+CD8+CD4−) | T cytotoxicCD8+ TNFa+ | % TNFa+ of CD8+ T cells | |
| ID51.T3 | (as of viable CD45+CD3+CD8+CD4−) | T cytotoxicCD8+ TNFa+ | % TNFa+ of CD8+ T cells | |
| ID52.T1 | (as of viable CD45+CD3+CD8+CD4−) | T cytotoxic CD8+ CD27+ | % CD27+ of CD8+ T cells | |
| ID52.T2 | (as of viable CD45+CD3+CD8+CD4−) | T cytotoxic CD8+ CD27+ | % CD27+ of CD8+ T cells | |
| ID52.T3 | (as of viable CD45+CD3+CD8+CD4−) | T cytotoxic CD8+ CD27+ | % CD27+ of CD8+ T cells | |

(continued)

**Table 3** (continued)

| Population ID | Parent population | Subset name | Name for paper | Comments |
|---|---|---|---|---|
| ID53 | (as of viable CD45+CD3+CD8+CD4−) | T cytotoxic CD8+ RA+ | % CD45RA+ of CD8+ T cells | |
| ID54 | (as of viable CD45+CD3+CD8+CD4−CD45RA+) | T cytotoxic Naïve | % CD27+ of CD45RA+ CD8+ T cells (Naïve CD8+T) | |
| ID55 | (as of viable CD45+ CD3+CD8+CD4−CD45RA+) | T cytotoxic EMRA | % CD27− of CD45RA+ CD8+ T cells (EMRA CD8+ T) | |
| ID56 | (as of viable CD45+CD3+CD8+CD4−) | T cytotoxic Total memory | % CD45RA− of CD8+ T cells (CD45RA−memory CD8+ T) | |
| ID57 | (as of viable CD45+CD3+CD8+CD4−CD45RA−) | T cytotoxic central memory | % CD27+ CCR7 − of CD45RA−memory CD8+ T cells (Central memory CD8+ T) | |
| ID58 | (as of viable CD45+CD3+CD8+CD4−CD45RA−) | T cytotoxic effector memory | % CD27−CCR7−of CD45RA− memory CD8+ T cells (Effector memory CD8+ T) | |
| ID59 | (as of viable CD45+CD3+CD4+CD8−) | Treg | % CD25hi FoxP3+ of CD4+ T cells (Treg) | |
| ID60 | (as of viable CD45+CD3+CD4+CD8−CD25hiFOXp3+) | Treg CD38+ | % CD38+ of Treg cells | |
| ID61 | (as of viable CD45+CD3+CD4+CD8−CD25hiFOXp3+) | Treg CD39+ | % CD39+of Treg cells | |
| ID62 | (as of viable CD45+CD3+CD4+CD8−CD25hiFOXp3+) | Treg CD103+ | % CD103+ of Treg cells | |
| ID63 | (as of viable CD45+CD3+CD4+CD8−CD25hiFOXp3+) | Treg HLA−DR+ | % HLA−DR+ of Treg cells | |
| ID64 | (as of viable CD45+) | Monocytes | % CD14+ of viable CD45+ cells (Total Monocytes) | |
| ID65 | (as of viable CD45+CD14+) | activated monocyte CD40+ | % CD40+ of total monocytes | |
| ID66 | (as of viable CD45+CD14+) | activated monocyte CD83+ | % CD83+ of total monocytes | |
| ID67 | (as of viable CD45+CD14+) | activated monocyte CD86+ | % CD86+ of total monocytes | |
| ID68 | (as of viable CD45+CD14+) | activated monocyte HLA−DR+ | % HLA−DR+ of total monocytes | |
| ID69 | (as of viable CD45+CD14+) | activated monocyteTNFa+ | % TNFa+ of total monocytes | |
| ID70 | (as of viable CD45+CD3−CD19−CD20−CD16−CD56−CD14−) | | Dendritic cells | % HLA−DR+ of Lin−CD45+ (Total Dendritic cells) |
| ID71 | (as of viable CD45+CD3−CD19−CD20−CD16−CD56−CD14−HLADR+) | | TNFa secreting DCs | % TNFa+ of total DCs |
| ID72 | (as of viable CD45+CD3−CD19−CD20−CD16−CD56−CD14−HLADR+) | | IFNa secreting DCs | % INFa+ of total DCs |
| ID73 | (as of viable CD45+CD3−CD19−CD20−CD16−CD56−CD14−HLADR+) | | Myeloid DC | % CD11c+ CD123− of total DCs (Myeloid DCs) |
| ID74 | (as of viable CD45+CD3−CD19−CD20−CD16−CD56−CD14−HLADR+CD11c+CD123−) | | TNFa secreting mDCs | % TNFa+ of mDCs |
| ID75 | (as of viable CD45+CD3−CD19−CD20−CD16−CD56−CD14−HLADR+ CD11c+CD123−) | | IFNa secreting mDCs | % IFNa+ of mDCs |
| ID76 | (as of viable CD45+CD3−CD19−CD20−CD16−CD56−CD14−HLADR+) | | Plasmacytoid dendritic cells | % CD11c−CD123+ of total DCs (Plasmacytoid DCs) |
| ID77 | (as of viable CD45+CD3−CD19−CD20−CD16−CD56−CD14−HLADR+CD11c−CD123+) | | TNFa secreting pDCs | % TNFa+ of pDCs |
| ID78 | (as of viable CD45+CD3−CD19−CD20−CD16−CD56− CD14−HLADR+ | | IFNa secreting pDCs | % INFa+ of pDCs |
| ID79 | (as of viable CD45+CD3−CD19−CD20−CD16−CD56− CD14−HLADR+) | | do not know this population | % CD11c+ CD123+ of total DCs |

(continued)

**Table 3** (continued)

| Population ID | Parent population | Subset name | Name for paper | Comments |
|---|---|---|---|---|
| ID80 | (as of viable CD45+) | Total B | % CD19+ of viable CD45+ (Total B cells) | |
| ID81 | (as of viable CD45+CD19+CD20+) | Mature Activated B CD80+ | % CD80+ of CD20+ B cells (CD80+ activated mature B) | |
| ID82 | (as of viable CD45+CD19+CD20+) | Mature Activated B CD86+ | % CD86+ of CD20+ B cells (CD86+ activated mature B) | |
| ID83 | (as of viable CD45+CD19+CD20+) | Mature B IgA+ | % IgA+ of CD20+ B cells (IgA+ mature B) | |
| ID84 | (as of viable CD45+CD19+CD20+) | Mature B IgG+ | % IgG+ of CD20+ B cells (IgG+ mature B) | |
| ID85 | (as of viable CD45+CD19+CD20+) | Mature B IgM+ IgD+ | % IgM+ IgD+ of CD20+ B cells (IgM+ IgD+ mature B) | |
| ID86 | (as of viable CD45+CD19+CD20+) | Mature B IgM−IgD− | % IgM−IgD−of CD20+ B cells (IgM−IgD− mature B) | |
| ID87 | (as of viable CD45+CD19+) | plasmablast | % CD27hi CD38hi of CD20− B cells (Plasmablasts) | |
| ID88 | (as of viable CD45+CD19+CD20−CD27highCD38++) | activated plasmablast CD23+ | % CD23+ of plasmablasts | |
| ID89 | (as of viable CD45+CD19+CD20−CD27highCD38++) | activated plasmablast CD21+ | % CD21+ of plasmablasts | |
| ID90 | (as of viable CD45+CD19 + CD20+) | Transitional B cells | % CD10+ CD27− of CD20+ B cells (Transitional B) | |
| ID91 | (as of viable CD45+CD19+CD20+CD27−CD10+) | Transitional B cell subset CD38+ | % CD38+ of transitional B cells | |
| ID92 | (as of viable CD45+CD19+CD20+CD27−CD10+) | Transitional B cell activated CD23+ | % CD23+ of transitional B cells | |
| ID93 | (as of viable CD45+CD19+CD20+CD27−CD10+) | Transitional B cell activated CD21+ | % CD21+ of transitional B cells | |
| ID94 | (as of viable CD45+CD19+CD20+ (minus viable CD45+CD19+CD27−CD10+)) | | Memory B cells IgD− | % IgD−CD27+ of CD20+ B cells* (IgD−CD27+ memory B) |
| ID95 | (as of viable CD45+CD19+CD20+IgD−CD27+ (minus viable CD45+CD19+CD27−CD10+) | | Memory B cells IgD−CD23+ | % CD23+ of IgD−CD27+ memory B cells |
| ID96 | (as of viable CD45+CD19+CD20+IgD−CD27+ (minus viable CD45+ CD19+CD27−CD10+) | | Memory B cells IgD−CD38+ | % CD38+ of IgD−CD27+ memory B cells** |
| ID97 | (as of viable CD45+CD19+CD20+IgD−CD27+ (minus viable CD45+CD19+CD27−CD10+) | | Memory B cells IgD−CD80+ | % CD80+ of IgD−CD27+ memory B cells |
| ID98 | (as of viable CD45+CD19+CD20+IgD−CD27+ (minus viable CD45+CD19+CD27−CD10+) | | Memory B cells IgD−CD86+ | % CD86+ of IgD−CD27+ memory B cells |
| ID99 | (as of viable CD45+CD19+CD20+IgD−CD27+ (minus viable CD45+CD19+CD27−CD10+) | | Memory B cells IgD− IgA+ | % IgA+ of IgD−CD27+ memory B cells |
| ID100 | (as of viable CD45+CD19+CD20+IgD−CD27+ (minus viable CD45+CD19+CD27−CD10+) | | Memory B cells IgD− IgG+ | % IgG+ of IgD−CD27+ memory B cells |
| ID101 | (as of viable CD45+CD19+CD20+ (minus viable CD45+CD19+CD27−CD10+) | | Memory B cells IgD+ | % IgD+ CD27+ of CD20+ B cells* (IgD+ CD27+ memory B) |
| ID102 | (as of viable CD45+CD19+CD20+IgD+CD27+ (minus viable CD45+CD19+ CD27−CD10+) | | Memory B cells IgD+CD23+ | % CD23+ of IgD+ CD27+ memory B cells |
| ID103 | (as of viable CD45+CD19+CD20+IgD+CD27+ (minus viable CD45+CD19+CD27−CD10+) | | Memory B cells IgD+CD38+ | % CD38+of IgD+CD27+ memory B cells*** |
| ID104 | (as of viable CD45+CD19+CD20+IgD+CD27+ (minus viable CD45+CD19+CD27−CD10+) | | Memory B cells IgD+ CD80+ | % CD80+ of IgD+ CD27+ memory B cells |
| ID105 | (as of viable CD45+CD19+CD20+IgD+CD27+(minus viable CD45+CD19+ CD27−CD10+) | | Memory B cells IgD+CD86+ | % CD86+ of IgD+CD27+ memory B cells |
| ID106 | (as of viable CD45+CD19+CD20+(minus viable CD45+CD19+CD27−CD10+) | | Naïve B cells | % IgD+ CD27− of CD20+ B cells* (Naïve B) |
| ID107 | (as of viable CD45+CD19+CD20+IgD+CD27− (minus viable CD45+CD19+CD27−CD10+) | | Naïve B cells activated CD21+ | % CD21+of Naïve B cells |

(continued)

**Table 3** (continued)

| Population ID | Parent population | Subset name | Name for paper | Comments |
|---|---|---|---|---|
| ID108 | (as of viable CD45+CD19+CD20+IgD+CD27− | (minus viable CD45+CD19+CD27−CD10+) | Naïve B cells activated CD38+ | % CD38+ of Naïve B cells |
| ID109 | (as of viable CD45+CD19+CD20+IgD+CD27− | (minus viable CD45+CD19+CD27−CD10+) | Naïve B cells activated CD80+ | % CD80+ of Naïve B cells |
| ID110 | (as of viable CD45+CD19+CD20+IgD+CD27− | (minus viable CD45+CD19+CD27−CD10+) | Naïve B cells activated CD86+ | % CD86+ of Naïve B cells |
| ID111 | (as of viable CD45+CD19+CD20+IgD+CD27− | (minus viable CD45+CD19+CD27−CD10+) | Naïve B cells IgA+ | % IgA+ of Naïve B cells |
| ID112 | (as of viable CD45+CD19+CD20+IgD+CD27− | (minus viable CD45+CD19+CD27−CD10+) | Naïve B cells IgG+ | % IgG+ of Naïve B cells |
| ID113 | (as of viable CD45+CD19+CD20+ | (minus viable CD45+CD19+CD27−CD10+) | Memory CD27−B cells | % IgD−CD27−of CD20+ B cells* (IgD−CD27−memory B) |
| ID114 | (as of viable CD45+CD19+CD20+IgD−CD27− | (minus viable CD45+CD19+CD27−CD10+) | Memory CD27−B cells CD21+ | % CD21+ of IgD−CD27−memory B cells |
| ID115 | (as of viable CD45+CD19+CD20+IgD−CD27− | (minus viable CD45+CD19+CD27−CD10+) | Memory CD27− B cells CD38+ | % CD38+ of IgD−CD27−memory B cells |
| ID116 | (as of viable CD45+CD19+CD20+IgD−CD27− | (minus viable CD45+CD19+CD27−CD10+) | Memory CD27−B cells CD80+ | % CD80+ of IgD−CD27−memory B cells |
| ID117 | (as of viable CD45+CD19+CD20+IgD−CD27− | (minus viable CD45+CD19+CD27−CD10+) | Memory CD27−B cells CD86+ | % CD86+ of IgD−CD27−memory B cells |
| ID118 | (as of viable CD45+CD19+CD20+IgD−CD27− | (minus viable CD45+CD19+CD27−CD10+) | Memory CD27−B cells CD23+ | % CD23+ of IgD−CD27−memory B cells |
| ID119 | (as of viable CD45+CD19+CD20+IgD−CD27− | (minus viable CD45+CD19+CD27−CD10+) | Memory CD27−B cells IgA+ | % IgA+ of IgD−CD27−memory B cells |
| ID120 | (as of viable CD45+CD19+CD20+IgD−CD27− | (minus viable CD45+CD19+CD27−CD10+) | Memory CD27−B cells IgG+ | % IgG+ of IgD−CD27−memory B cells |

A number of computational approaches have been published for automation of the analysis of flow cytometric data and for better understanding and presentation of high-dimensional data (Pyne et al. 2009; Rogers and Holyst 2009; Qian et al. 2010; Bagwell 2011; Dabdoub et al. 2011; Qiu et al. 2011; Aghaeepour et al. 2012; Candia in press). Most notable are the efforts of FlowCAP (Flow Cytometry Critical Assessment of Population Identification Methods), a consortium of investigators working together to advance the development of computational models for the analysis of flow cytometry data (Aghaeepour et al. 2013; Biehl et al. 2013). Such approaches are necessary not only for data mining of high-dimensional data, but also to remove the subjectivity inherent to manual gating to increase the reproducibility of these complex assays. As is discussed in a separate chapter of this edition, these approaches for automated analysis are rapidly evolving and show great promise. One recent study, using data derived during the development of our CLIP panel, demonstrated the ability to accurately classify two autoimmune diseases from a 16-parameter (14-color) immunophenotyping of peripheral blood from these patients (Candia in press). Perhaps the biggest obstacle to development or acceptance of these automated approaches is the need to compare them during the development phase to existing 'gold standards' for routine analyses. Unfortunately, the gold standard in this instance is manual gating of cellular populations—a process that is fraught with subjectivity, and hence great variation.

## 4 Future Directions and Summary

Clearly, any attempt to immunophenotype the entire immunome simultaneously remains in its infancy. One can think of nearly endless possibilities to better characterize the immune system, including pursuits such as studying signaling pathways in various lineages in response to stimuli. Furthermore, new markers are continually being discovered that identify new cell subpopulations or that correlate with specific cellular functions. Thus, immunophenotyping of the immune will be a constantly evolving process due to both biological as well as technical innovations. Nonetheless, it is important to begin efforts toward this pan-immunophenotyping as these data will prove to be a valuable resource, and lessons learned from these early steps will help to guide future efforts.

Since the inception of our CLIP panel, modifications made for biological and technical reasons. Our initial panel relied on the use of quantum dot (QDot) conjugates for three antibodies (QD605, QD655, and QD800) in each tube. QDots permitted substantial use of the violet laser (406 nm) for increasing the number of detectable parameters in the CLIP panel and are widely used in multicolor cytometry. They are generally more stable than tandem conjugates, allowing good reproducibility. Unfortunately, QDot conjugates containing unbound particles can interact with each other to form aggregates—a problem that increases as the reagents age. These aggregates appear with dual fluorescence from each QDot

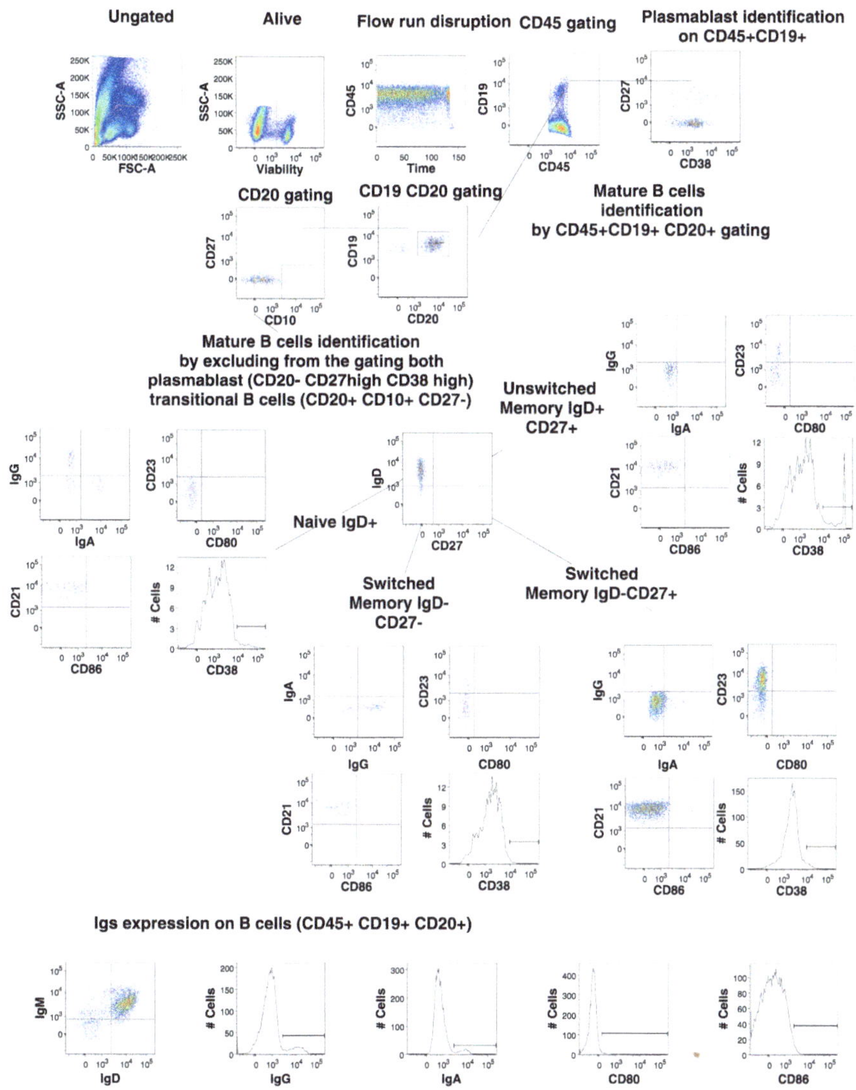


**Fig. 9** Example of gating procedure for B cell populations identified by CLIP. Single-cell suspensions from healthy donors were stained with a combination of viability dye and 14 antibodies from our B2 staining panel. Lymphocytes were identified based on their FSC-SSC properties. Dead cells were excluded through the use of a viability dye. CD45 and CD19 were used to identify B cells ($CD45^{+}CD19^{+}$) among the previously selected living lymphocytes. This figure illustrates the gating strategy that allowed us to identify 40 B cell populations

spectrum and can make determination of proper compensation extremely difficult. The presence of these aggregates can be minimized by subjecting the antibody dilutions to a quick high-speed spin, preventing the non-specific labeling of some

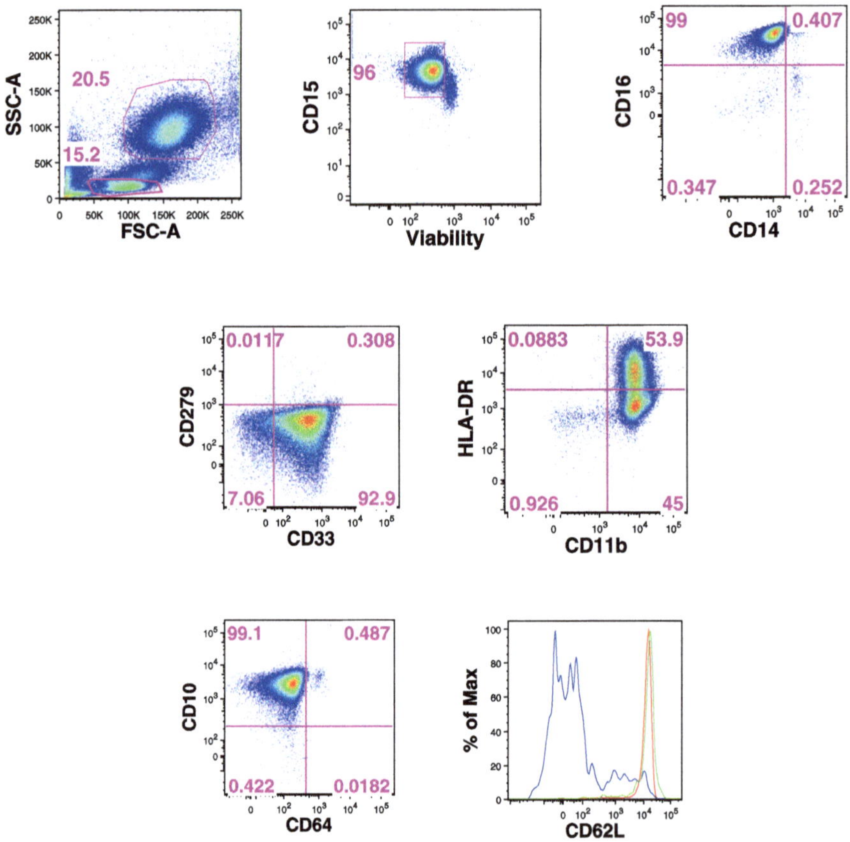


**Fig. 10** Example of the gating procedure for neutrophils identified by CLIP. Ammonium chloride-lysed single-cell suspensions from healthy donors were stained with a combination of viability dye and 14 antibodies from our N1 staining panel. Neutrophils were identified based on their FSC-SSC properties and CD15 was used as a lineage marker. Mature or immature neutrophils were identified by CD10 versus CD64 expression, and overlay of the expression of the homing marker CD62L identified mature $CD10^{+}CD64^{-}$ (*red line*), mature activated $CD10^{+}CD64^{+}$ (*green line*), and immature $CD10^{-}CD64^{-}$ (*blue line*) neutrophils

cells (Chattopadhyay et al. 2006, 2010), or by gating out obvious aggregates. The new generation of Brilliant Violet (BV) stains ($\pi$-conjugated polymers) has proven to be a practical alternative to QDots as BV stains have high quantum yields, good stability, and are available in emission spectra similar to those of QDots (Chattopadhyay et al. 2012). BV-conjugated antibodies have now replaced the QDots in our CLIP panel and have completely eliminated compensation problems that arose from QDots aggregates.

Changes have also been made to the original panel based on marker discovery. For example, CD146 (MCAM) was recently identified as a marker of committed Th17 cells in humans. CD146 was introduced into the Th17 tube as a replacement for IL-21, which had produced rather uninformative data (Dagur et al. 2011). CD146 appears to be involved in the extravasation and migration of these cells, and data suggest that Th17 cells are important in a number of autoimmune diseases (Amadi-Obi et al. 2007; Kebir et al. 2007; Brucklacher-Waldert et al. 2009; Kleinschek et al. 2009; Dagur et al. 2011). In one tube CD45-RA was replaced with CD45-RO because of dim staining of the former when used in conjunction with a specific fixation and permeabilization procedure. Additional changes are also under study.

It is clear that the comprehensiveness of our current panel is limited and does not begin to cover the entire immunome. A vast array of additional features and markers could be studied, and there remains a need to add tubes designed to study additional cell types such as T follicular helper cells and peripheral blood progenitor cells. Our approach has been to design and implement one tube at a time and then to integrate it into the larger CLIP panel after validation. Development is anticipated to continue in this manner. It is also entirely possible that the design of the CLIP panel could be changed in order to add more parameters. Hardware improvements such as the spectral flow cytometry together with the development of novel fluorochromes with tight emission spectra would permit the practical examination of more fluorescent probes in a single tube and thus make this panel more comprehensive.

The original CLIP panel was designed for use on specimens from clinical trials, the overwhelming majority of which are cryopreserved and thus contain only mononuclear cells. Thus the initial panel is focused on mononuclear cell subsets to the exclusion of others. More recently, the CLIP panel has become truly more comprehensive by inclusion of two tubes designed specifically for granulocytic cells. Expansion of the CLIP assay to whole blood is necessary but will be accompanied by the need to perform studies 'on demand'. This consideration, together with the need for a high degree of reproducibility over time in these studies, makes a strong argument against the manual preparation of the antibody cocktails each time the CLIP phenotyping is to be performed. To circumvent this necessity, there are three options: (1) to make use of automated devices for antibody dispensing, (2) to make large batches of antibody cocktails, and (3) to lyophilized pre-titered antibody cocktails. As automated liquid handlers tend to be expensive, require laboratory space, and are subject to period breakdown, the first approach would be our least favored alternative. The most attractive option is lyophilization of the panel. This approach has been pioneered as a means of widespread standardization of immunophenotyping for clinical trial samples and offers significant advantages in terms of reagent stability, storage, and ease of use (Maecker et al. 2012). A previous attempt to lyophilize our CLIP panel did not prove successful, exclusively, due to the failure of the QDots to maintain their performance characteristics after lyophilization. The replacement of QDots with the BV stains in the current iteration of our panel may make lyophilization of these reagents feasible. A remaining obstacle to this process is the intracellular staining of cytokines that requires cells to be fixed and permeabilized prior to staining. To address this, either the lyophilized reagent cocktail would need to be limited to only the cell surface markers or methods would need to be devised for permeabilization of the cells prior to adding them to the antibody cocktail.

As mentioned previously, one of the tremendous advantages of using fluorescence-based high-dimensional immunophenotyping rather than the time-of-flight cytometry is the ability to recover any subsets identified for function analyses. As discussed above, the current CLIP panel can be used to purify most subsets identified, although modifications may be needed for those populations identified by secretion of cytokines (Fitzgerald and Grivel 2013). With the imminent possibilities for lyophilization of pre-titered, premixed reagents and computational methods for setting compensation as well as for data analysis, performance of high polychromatic flow cytometry will become a practical approach for the comprehensive immunophenotyping of the immune system. In addition to permitting a rapid, in-depth study of the immunome, this approach should also prove to be robust with low variance over time. Such work, when combined with ex vivo functional studies of novel populations and other data such as gene expression studies, antibody titers, Elispot data, serum cytokine analysis, epigenetic studies, and proteomic stand to revolutionize our understanding of human immunology.

## References


Aghaeepour N, Finak G, Flow CAPC, Consortium D, Hoos H, Mosmann TR, Brinkman R, Gottardo R, Scheuermann RH (2013) Critical assessment of automated flow cytometry data analysis techniques. Nat Methods 10:228–238

Aghaeepour N, Jalali A, O'Neill K, Chattopadhyay PK, Roederer M, Hoos HH, Brinkman RR (2012) RchyOptimyx: cellular hierarchy optimization for flow cytometry. Cytometry A 81:1022–1030

Amadi-Obi A, Yu CR, Liu X, Mahdi RM, Clarke GL, Nussenblatt RB, Gery I, Lee YS, Egwuagu CE (2007) TH17 cells contribute to uveitis and scleritis and are expanded by IL-2 and inhibited by IL-27/STAT1. Nat medicine 13:711–718

Bagwell CB (2011) Breaking the dimensionality barrier. Methods Mol Biol 699:31–51

Bendall SC, Nolan GP, Roederer M, Chattopadhyay PK (2012) A deep profiler's guide to cytometry. Trends Immunol 33:323–332

Beyer M, Kochanek M, Darabi K, Popov A, Jensen M, Endl E, Knolle PA, Thomas RK, von Bergwelt-Baildon M, Debey S, Hallek M, Schultze JL (2005) Reduced frequencies and suppressive function of CD4 + CD25hi regulatory T cells in patients with chronic lymphocytic leukemia after therapy with fludarabine. Blood 106:2018–2025

Biancotto A, Dagur PK, Fuchs JC, Langweiler M, McCoy JP Jr (2012a) OMIP-004: in-depth characterization of human T regulatory cells. Cytometry A 81:15–16

Biancotto A, Dagur PK, Fuchs JC, Wiestner A, Bagwell CB, McCoy JP Jr (2012b) Phenotypic complexity of T regulatory subsets in patients with B-chronic lymphocytic leukemia. Mod Pathol 25:246–259

Biancotto A, Fuchs JC, Williams A, Dagur PK, McCoy JP Jr (2011) High dimensional flow cytometry for comprehensive leukocyte immunophenotyping (CLIP) in translational research. J Immunol Methods 363:245–261

Biehl M, Bunte K, Schneider P (2013) Analysis of flow cytometry data by matrix relevance learning vector quantization. PLoS ONE 8:e59401

Brucklacher-Waldert V, Stuerner K, Kolster M, Wolthausen J, Tolosa E (2009) Phenotypical and functional characterization of T helper 17 cells in multiple sclerosis. Brain 132:3329–3341

Candia JMR, Driscoll M, Biancotto A, Dagur P, McCoy JP, Sen HN, Wei L, Maritan A, Cao K, Nussenblatt RB, Banavar JR, Losert W (in press) From cellular characteristics to disease diagnosis: uncovering phenotypes with supercells. PLOS Comput Biol. doi: 10.1371/journal.pcbi.1003215

Chattopadhyay PK, Gaylord B, Palmer A, Jiang N, Raven MA, Lewis G, Reuter MA, Nur-ur Rahman AK, Price DA, Betts MR, Roederer M (2012) Brilliant violet fluorophores: a new class of ultrabright fluorescent compounds for immunofluorescence experiments. Cytometry A 81:456–466

Chattopadhyay PK, Perfetto SP, Yu J, Roederer M (2010) The use of quantum dot nanocrystals in multicolor flow cytometry. Nanomed nanobiotechnol 2:334–348

Chattopadhyay PK, Price DA, Harper TF, Betts MR, Yu J, Gostick E, Perfetto SP, Goepfert P, Koup RA, De Rosa SC, Bruchez MP, Roederer M (2006) Quantum dot semiconductor nanocrystals for immunophenotyping by polychromatic flow cytometry. Nat Medicine 12:972–977

D'Arena G, Laurenti L, Minervini MM, Deaglio S, Bonello L, De Martino L, De Padua L, Savino L, Tarnani M, De Feo V, Cascavilla N (2011) Regulatory T-cell number is increased in chronic lymphocytic leukemia patients and correlates with progressive disease. Leuk Res 35:363–368

Dabdoub SM, Ray WC, Justice SS (2011) FIND: a new software tool and development platform for enhanced multicolor flow analysis. BMC Bioinformatics 12:145

Dagur PK, Biancotto A, Wei L, Sen HN, Yao M, Strober W, Nussenblatt RB, McCoy JP Jr (2011) MCAM-expressing CD4(+) T cells in peripheral blood secrete IL-17A and are significantly elevated in inflammatory autoimmune diseases. J Autoimmun 37:319–327

Davis MM (2008) A prescription for human immunology. Immunity 29:835–838

Davis MM (2012) Immunology taught by humans. Sci transl med 4:117fs112

Deutsch V, Perry C, Polliack A (2009) Expansion of regulatory T cells in B chronic lymphocytic leukemia: enhanced 'brakes' on host immunity. Leuk Lymphoma 50:687–688

Dickler HB, McCoy JP, Nussenblatt R, Perl S, Schwartzberg PA, Tsang JS, Wang E, Young NS (2013) The national institutes of health center for human immunology, autoimmunity, and inflammation: history and progress. Ann N Y Acad Sci 1285:133–147

Fitzgerald W, Grivel JC (2013) A universal nanoparticle cell secretion capture assay. Cytometry A 83:205–211

Giannopoulos K, Schmitt M, Kowal M, Wlasiuk P, Bojarska-Junak A, Chen J, Rolinski J, Dmoszynska A (2008) Characterization of regulatory T cells in patients with B-cell chronic lymphocytic leukemia. Oncol Rep 20:677–682

Jak M, Mous R, Remmerswaal EB, Spijker R, Jaspers A, Yague A, Eldering E, Van Lier RA, Van Oers MH (2009) Enhanced formation and survival of CD4 + CD25hi Foxp3 + T-cells in chronic lymphocytic leukemia. Leuk Lymphoma 50:788–801

Kebir H, Kreymborg K, Ifergan I, Dodelet-Devillers A, Cayrol R, Bernard M, Giuliani F, Arbour N, Becher B, Prat A (2007) Human TH17 lymphocytes promote blood-brain barrier disruption and central nervous system inflammation. Nat Medicine 13:1173–1175

Kleinschek MA, Boniface K, Sadekova S, Grein J, Murphy EE, Turner SP, Raskin L, Desai B, Faubion WA, de Waal Malefyt R, Pierce RH, McClanahan T, Kastelein RA (2009) Circulating and gut-resident human Th17 cells express CD161 and promote intestinal inflammation. J Exp Med 206:525–534

Maecker HT, McCoy JP, Nussenblatt R (2012) Standardizing immunophenotyping for the Human Immunology Project. Nat Rev Immunol 12:191–200

Mahnke YD, Roederer M (2007) Optimizing a multicolor immunophenotyping assay. Clin lab med 27:469–485

McCoy JP Jr, Overton WR (1994) Quality control in flow cytometry for diagnostic pathology: II. a conspectus of reference ranges for lymphocyte immunophenotyping. Cytometry 18:129–139

Perfetto SP, Ambrozak D, Nguyen R, Chattopadhyay P, Roederer M (2006) Quality assurance for polychromatic flow cytometry. Nat Protocols 1:1522–1530

Perfetto SP, Ambrozak D, Nguyen R, Chattopadhyay PK, Roederer M (2012) Quality assurance for polychromatic flow cytometry using a suite of calibration beads. Nat Protocols 7:2067–2079

Perfetto SP, Chattopadhyay PK, Lamoreaux L, Nguyen R, Ambrozak D, Koup RA, Roederer M (2010) Amine-reactive dyes for dead cell discrimination in fixed samples. Curr Protoc Cytom (Chapter 9, Unit 9.34)

Perfetto SP, Chattopadhyay PK, Roederer M (2004) Seventeen-colour flow cytometry: unravelling the immune system. Nat Rev Immunol 4:648–655

Pyne S, Hu X, Wang K, Rossin E, Lin TI, Maier LM, Baecher-Allan C, McLachlan GJ, Tamayo P, Hafler DA, De Jager PL, Mesirov JP (2009) Automated high-dimensional flow cytometric data analysis. PNAS 106:8519–8524

Qian Y, Wei C, Eun-Hyung Lee F, Campbell J, Halliley J, Lee JA, Cai J, Kong YM, Sadat E, Thomson E, Dunn P, Seegmiller AC, Karandikar NJ, Tipton CM, Mosmann T, Sanz I, Scheuermann RH (2010) Elucidation of seventeen human peripheral blood B-cell subsets and quantification of the tetanus response using a density-based method for the automated identification of cell populations in multidimensional flow cytometry data. Cytometry B Clin cytom 78(1):S69–82

Qiu P, Simonds EF, Bendall SC, Gibbs KD Jr, Bruggner RV, Linderman MD, Sachs K, Nolan GP, Plevritis SK (2011) Extracting a cellular hierarchy from high-dimensional cytometry data with SPADE. Nat Biotechnology 29:886–891

Roederer M (2002) Compensation in flow cytometry. Curr protoc cytom (Chapter 1, Unit 1.14)

Rogers WT, Holyst HA (2009) FlowFP: a bioconductor package for fingerprinting flow cytometric data. Adv bioinform 193947:11

Sony (2013) Sony announces the development of the 'Spectral' cell analyzer. sony. http://www.sony.net/SonyInfo/News/Press/201206/12-082E/index.html

Tsang JS, Schwartzberg PL, Kotliarov Y, Biancotto A, Xie Z, Wang EMO, Narayanan M, Cheung F, Golding H, Moir S, Ho J, Khurana S, Shum P, Perl S, Fuchs JC, Langweiler M, Nussenblatt R, Kastner D, Trinchieri G, Marincola F, Young NS, Dickler H, McCoy JP, Germain RN, JS T (2013) Global analyses of human immune states utilizing natural variations and vaccine perturbation reveals baseline predictors of the post-vaccination response. Cell (submitted)

# High-Dimensional Analysis of Human CD8$^+$ T Cell Phenotype, Function, and Antigen Specificity

**Evan W. Newell and Wenyu Lin**

**Abstract** Antigen-specific T cells are critical initiators and orchestrators of the adaptive immune response. Categorizing antigen-specific T cell subsets is not a simple task given the diversity of these cells and the large number of parameters that can be considered. Here, we focus on human CD8$^+$ T cells and discuss the utility of high-dimensional mass cytometric analysis techniques for the concurrent identification and characterization of antigen-specific T cells involved in immunological homeostasis and disease. We first provide an overview of previously identified T cell subsets. We then discuss the segregation of antigen-specific T cells based on protein expression through surface and/or intracellular staining, on functional capacity through measurement expression of cytokines or other inducible markers, and on the antigen-specificity of the cell assessed using peptide-major histocompatibility complex multimers. High-dimensional mass cytometry enables a deeper and more integrated view of all three aspects of antigen-specific T cell diversity than do traditional techniques. Use of mass cytometry for precise measurement of the status of antigen-specific immune responses should result in better prediction of vaccine efficacy and disease outcomes.

## Contents



E. W. Newell (✉) · W. Lin
Singapore Immunology Network (SIgN), Agency for Science Technology and Research (A-STAR), Singapore 138648, Republic of Singapore
e-mail: evan_newell@immunol.a-star.edu.sg

Current Topics in Microbiology and Immunology (2014) 377: 61–84
DOI: 10.1007/82_2013_354


Published Online: 6 October 2013

# 1 Introduction

T cells are critical regulators of adaptive immune responses and are involved in a wide range of human diseases. As such, understanding the nature of the T cell response by identifying T cell correlates of human immune response outcomes in vaccination, infection, or other diseases will improve therapeutic management. Here, we discuss strategies for assessing both quantitative and qualitative features of the T cell response. In terms of quantity, we will consider the breadth and magnitude of the antigen-specific response. In terms of quality, we will focus on understanding T cell differentiation and the associated changes in cellular phenotype and functional capacity. The assessment of T cell response has already been successful for a range of applications including observations about the relationship between the multi-functionality and efficacy of the T cell-mediated immune response (Betts et al. 2006; Seder et al. 2008; Yuan et al. 2008). By increasing the breadth of antigen specificities that can be probed simultaneously with a high-dimensional assessment of T cell phenotype and functions using mass cytometry, we hope to increase the power to define the status of a given T cell response with more precision.

T cells are one of the most diverse cell types in the body. Diversity in T cell receptor-mediated antigen specificity arises from inexact and random gene recombination. Antigen specificity aside, methodological progress in the study of T cell phenotypic and functional properties is leading to a seemingly endless number of ways to subdivide T cells, although the level at which these subdivisions are accurate and meaningful are not clear. In this review, we will discuss the utility of a high-dimensional mass cytometry analysis for identifying meaningful relationships among phenotype, function, and antigen specificity. We restrict our focus to well-studied and classically defined major histocompatibility complex (MHC) class I-restricted $\alpha\beta$ T cell receptor (TCR) expressing $CD8^+$ cytotoxic T cells; these cells are critical for the clearance of a wide range of viral infections and tumors. We expect that this type of analysis will be applicable for delineating other T cell subsets such as $CD4^+$ T cells, $\gamma\delta$ T cells, natural killer T cells, invariant T cells (such as mucosa-associated invariant T or MAIT cells), and other new 'non-classical' T cell subsets with phenotypes similar to those of the classical ones (Le Bourhis et al. 2013; Van Rhijn et al. 2013).

## 2 Classifying Human T Cells

Classically MHC-restricted $\alpha\beta$ T cells are derived from hematopoietic lymphoid progenitors and express a heterodimeric TCR, whose alpha and beta chains are each produced by a complicated process of *V(D)J* gene recombination. The end result of gene recombination is that most T cells express a single TCR but that these cells have enormous diversity (there are more than $10^{14}$ possible TCRs) (Davis and Bjorkman 1988). Through intrinsic (germline-encoded) bias (Feng et al. 2007; Garcia et al. 2009; Scott-Browne et al. 2009) and by thymic positive and negative selection, mature T cells express receptors that have very weak affinity for self-peptide antigens presented by self-MHC (Starr et al. 2003). The same cells, through higher affinity interactions, are also able to respond to a very small subset of the enormously large number of possible foreign (or self in the case of an auto-reactive T cell) peptides presented by the same MHC allele. In general, for any given foreign peptide bound to MHC, about 1 in 10,000–1 in 1 million T cells will react to that peptide-MHC. Because the number of possible peptides in the universe of peptides that can bind a given MHC is much larger, it is certain that each T cell must be able to respond degenerately to a large number of different peptide antigens (Holler and Kranz 2004). Nonetheless, each T cell is exquisitely specific for a tiny fraction of these possible peptides, making them essentially "epitope-specific." Due to the very large number of possible 'specificities' ($\sim$10,000–1 million), the working diversity of T (and B) cells of the adaptive immune system are unmatched by any other cell type. This makes them challenging, but appropriate, targets for high-dimensional analysis approaches.

An appreciation of the phenotypic and functional complexity of the T cell compartment is increasing. In humans, differential expression of CD45 isoforms by primed or memory human lymphocytes was first observed in 1988 (Akbar et al. 1988; Sanders et al. 1988). Later, additional types of memory T cells were identified through analysis of CD45 isoform expression in conjunction with CD27 (Hamann et al. 1997), CCR7, or CD62L expression (Sallusto et al. 1999). Based on proliferative potential, cytotoxicity capacity, and the immediate ability to produce cytokines, four major subsets can be distinguished: naïve ($T_N$, CCR7$^+$CD45RA$^+$), central memory ($T_{CM}$, CCR7$^+$CD45RA$^-$), effector memory ($T_{EM}$, CCR7$^-$CD45RA$^-$), and terminal effector ($T_{EMRA}$, CCR7$^-$CD45RA$^+$) (Sallusto et al. 1999). Compared to $T_{CM}$ cells, CD8$^+$ $T_{EM}$ cells exhibit rapid effector function, carry large amounts of perforin, and produce IFN-$\gamma$, within hours after antigenic stimulation (Sallusto et al. 2004). The use of these subset definitions is particularly attractive because the same receptors (CCR7 and CD62L) are involved in lymph node trafficking, giving a meaningful distinction between central memory cells in the blood en route to a lymph node and effector memory cells that are recirculating to non-lymphoid tissues (Masopust and Schenkel 2013).

Although definitions of the four major subsets of human T cells are useful and well accepted, it is clear that much more heterogeneity exists than are captured by these subdivisions and that these definitions are not always accurate. For instance, combinatorial expression of CD27, CD62L, and CCR7 can be considered, leading to a huge number of possible subsets. Furthermore, segregation of effector and memory T cell subsets can be accomplished by the inclusion of other markers, such as CD28, a co-stimulatory molecule with reduced expression in senescent effector cells (Romero et al. 2007). Despite this, even the simple distinction of naïve versus memory status is not a trivial one. A subset of cells with higher levels of CD95, IL-2Rb, CXCR3, and LFA-1 that otherwise fit a very strict definition of naïve cells ($CD45RA^+$, $CD45RO^-$, $CD27^+$, $CCR7^+$, $CD62L^+$, $CD28^+$) display characteristic of stem cell-like memory cells with high proliferative capacity (Gattinoni et al. 2011) (Table 1). This is highlighted further by the notion that homeostatic proliferation can lead to 'antigen-naïve' cells with memory phenotypes and functional capacities (Sprent and Surh 2011) and the recent description of cells specific for unexposed antigens that express memory markers (presumably cross-reactive with other antigens) and that have elevated functional capacities (Su et al. 2013).

We propose that by including a large number of phenotypic and functional markers in the analysis, we can come closer to true signatures of T cell status. With the aid of mass cytometry, concurrent analysis of more than 30 parameters is possible. Subsequently, a dimensionality reduction approach using principal component analysis (PCA) was able to delineate cells with naïve-like versus. memory-like properties (Newell et al. 2012). Although this approach is limited by the choice of markers and functional outcomes selected for the characterization of T cells, our hope is that the combination of information gained from all markers (including functional measures, transcription factors, and other signaling molecules) will allow for a segregation of naïve versus antigen-experienced cell subsets in a more meaningful way.

Much effort has been dedicated to understanding fate decisions made by responding T cells after antigen encounter. Some T cells participate only as effectors, whereas others are destined to become long-lived memory cells. Based on differential gene expression profiles (Kaech et al. 2002) and studies of IL-7R (CD127) expression (Schluns et al. 2000), IL-7R expressing effector cells were found to be predisposed to become long-lived memory cells during lymphocytic choriomeningitis virus or other infections in mice. These cells are termed memory precursor effector cells (MPEC). In contrast, cells with reduced expression of IL-7R, CD28, and CD27 but enhanced expression of CD57 and KLRG1 are senescent and have reduced telomere lengths, proliferative capacity, and survival; these are short-lived effector cells (SLEC) (Kaech et al. 2003). The exact determinant of fate decision is not totally clear, but depends on the level and duration of inflammation present during priming (Joshi et al. 2007), reciprocal expression of various transcription factors (e.g., T-bet, eomesodermin, Blimp1, Bcl-6), and the extent of antigen–receptor stimulation (Kaech and Cui 2012). Recent studies using genetically encoded barcode technologies for simultaneously tracking of large numbers of individual cells shows a great deal of variability in the magnitude and phenotypes of individual responding T cells that express identical TCRs. In terms of memory versus effector fates, it appears that a single cell can give rise to both, but that the proportion is skewed stochastically between families. This hypothesis

**Table 1** An evolving list of CD8+ T cell subsets and their definitions.

| Subset name | Definition | Surface marker expression | Functional capacities | Trafficking properties | Example antigen specificities |
|---|---|---|---|---|---|
| Naïve ($T_N$) | No specific antigen exposure | $CD45RA^+$ $CD45RO^-$ $CD27^+$ $CCR7^+$ $CD62L^+$ | Very little immediate cytokine production, survey APC for foreign antigens | Circulate between blood and lymphoid organs | Altered Mart-1 (Anichini et al. 2003) |
| Stem cell memory ($T_{SCM}$) ) | Stem cell-like memory T cells with high proliferative capacity and multipotency | $CD45RA^+$ $CD45RO^-$$CD27^+$ $CCR7^+$ $CD62L^+$ $CD95^+$ $IL2Rbeta^+$ $CD127^+$ $CXCR3^+$ $LFA\text{-}1^+$ | Can produce IFN-$\gamma$ only, not IL-2 or TNF. Differentiate into $T_{CM}$, $T_{EM}$, and $T_{Eff}$ in vitro | Presumably to lymph nodes | Influenza, CMV, Mart-1 in melanoma patients (Gattinoni et al. 2011) |
| Central memory ($T_{CM}$) | Long-lived memory cells; initial definition based on $CD45RO^+$ memory cells that express CCR7 and CD62L | $CD45RA^-$ $CD45RO^+$ CCR7 $CD127^{hi}$ $CD62L^{hi}$ (Sallusto et al. 1999) | $IL\text{-}2^+$, antigen-primed cells that lack immediate effector function, differentiate into effector cells upon secondary stimulation | Circulate between blood and lymphoid organs | Influenza, EBV (Ravkov et al. 2003), rotavirus (Newell et al. 2013) |
| Effector memory ($T_{EM}$) | Memory cells that have lost the expression of CCR7 and display chemokine receptors and adhesion molecules that are required for homing to inflamed tissues (Sallusto et al. 1999) | $CD45RA^-$ $CCR7^-$ (Sallusto et al. 1999) | Antigen-primed cells that can enter peripheral tissues rapidly to mediate inflammation, weakly proliferative (Masopust et al. 2001) | Migrate to inflamed peripheral tissues (Sallusto et al. 1999) | EBV (Appay et al. 2002), rotavirus (Newell et al. 2013), CMV (Gillespie et al. 2000) |

(continued)

**Table 1** (continued)

| Subset name | Definition | Surface marker expression | Functional capacities | Trafficking properties | Example antigen specificities |
|---|---|---|---|---|---|
| Recirculating memory ($T_{RCM}$) | Tissue recirculating effector memory cells | CD45RA$^-$ CD45RO$^+$ CCR7$^+$ CD62L$^{+/-}$CD103$^{+/-}$ | Restrict pathogen replication and spread before secondary effector cells generated (Mueller et al. 2013) | Recirculate between lymphoid and non-lymphoid tissues through expression of CCR7, CD103, and other trafficking receptors | Lung airways (Ely et al. 2006) |
| Resident memory ($T_{RM}$) | Permanent memory T cells that reside in peripheral tissues after an infection is cleared (Gebhardt et al. 2009; Masopust et al. 2010) | CD45RA$^-$ CCR7$^-$ CD27$^-$GranzymeB$^{lo}$ CD103$^+$ CD69$^+$ (Gebhardt et al. 2009; Masopust et al. 2010) | Provide site-specific protection; terminally differentiated (Wakim et al. 2010) | Do not traffic (Gebhardt et al. 2009; Masopust et al. 2010) | Rotavirus (Newell et al. 2013), herpes simplex virus (Zhu et al. 2009) |
| Early activated effector ($T_{act}$) | Newly generated effector T cells | CD38$^+$ HLA$^-$ DR$^+$ Ki67 Bcl-2$^+$ CD127$^-$ KLRG1$^-$ T-bet$^+$ perforin$^+$ (Miller et al. 2008; Anichini et al. 2010) | Very recent antigen exposure and TCR triggering | Trafficking receptors for affected tissues | Small pox (Miller et al. 2008), yellow fever (Miller et al. 2008), influenza (Wilkinson et al. 2012), celiac disease unknown target (Han et al. 2013) |
| Effector ($T_{Eff}$) | Effector functions such as the production of cytokines and lysis of pathogen-infected cells in a TCR-dependent manner (Nolz et al. 2011) | CD45RA$^-$ CCR7$^-$GranzymeB$^+$ KLRG-1$^+$ CD62L$^-$ CD127$^-$ | Short-lived, production of cytokines, lysis of pathogen-infected cells | Rapidly localize to areas of inflammation in order to fight infection (Agace 2006) | EBV, CMV |

(continued)

**Table 1** (continued)

| Subset name | Definition | Surface marker expression | Functional capacities | Trafficking properties | Example antigen specificities |
|---|---|---|---|---|---|
| Terminally differentiated effector ($T_{EMRA}$) | Highly differentiated T cells re-expressing CD45RA (Henson et al. 2012) | CD45RA$^+$ CCR7$^-$ and CD27$^-$ or CD28$^-$ (Saule et al. 2006) | | Preferential localization to peripheral and inflamed tissues and not the lymph nodes | CMV (Derhovanessian et al. 2011) |
| Senescent | Proliferative; dysfunction unresponsive/ terminally differentiated; telomeres shortened | CD45RA$^+$ CCR7$^-$ CD27$^-$CD28$^-$ and Tim-3$^+$ (Li et al. 2012), CD57$^+$ (Brenchley et al. 2003), KLRG-1$^+$ (Heffner and Fearon 2007) | Defective killing ability and negative regulatory functions (Appay et al. 2000; Akbar and Henson 2011) | | CMV (Pawelec et al. 2006) |
| Exhausted | Dysfunctional cytokine production and effector function due to repeated stimulation, resistant to reactivation (Wherry 2011) | PD-1$^+$ (Barber et al. 2006), Tim-3$^+$ (Sakuishi et al. 2010), LAG-3$^+$ (Blackburn et al. 2009), BTLA$^+$ (Derre et al. 2010) CD160$^+$ (Peretz et al. 2012), KLRG-1$^+$ (Bengsch et al. 2010), 2B4$^+$ (Bengsch et al. 2010) | Impaired IL-2, TNF-$\alpha$, IFN-$\gamma$ production, loss of proliferative capacity; loss of ex vivo killing capability; deletion of virus-specific T cells (Wherry et al. 2003; Wherry 2011) | | Hepatitis C virus (HCV) (Golden-Mason et al. 2009), human immunodeficiency virus (HIV) (Jones et al. 2008) |

(continued)

**Table 1** (continued)

| Subset name | Definition | Surface marker expression | Functional capacities | Trafficking properties | Example antigen specificities |
|---|---|---|---|---|---|
| Anergic | Induced hyporesponsive state when T cells are stimulated with low co-stimulatory and/or high co-inhibitory signaling (Crespo et al. 2013a) | Poorly defined in humans | Low IL-2 production, cell cycle arrest at G1/S phase (Crespo et al. 2013a) | | |
| Invariant T cells | | | | | |
| Mucosal associated invariant T cells (MAIT) | T cells that possess a semi-invariant TCR and are restricted by the evolutionarily conserved major histocompatibility complex related molecule, MR1 (Le Bourhis et al. 2011) | $CD161^+$, $Va7.2^+$, IL-$18R\alpha^+$ (cord blood), effector ($CD95^{hi}$ $CD62L^{lo}$) memory ($CD27^+$ $CD45R0^+$ $CD45RA^{lo}$ $CD122^+$) phenotype; mostly $CD8\alpha\beta^{lo}$ or DN cells with few CD4 single-positive cells (Le Bourhis et al. 2011) | Detect intracellular infection of non-professional antigen presenting cells; provide early and rapid control of intracellular pathogens (Gold and Lewinsohn 2011), provide an early source of IFN-$\gamma$ that enhances Th1 responses (Gold and Lewinsohn 2011) | Enriched in non-lymphoid organs but present at low frequencies in lymph nodes (Gold and Lewinsohn 2011) $CCR6^+$ $CXCR6^+$ $CCR7^-$ (Le Bourhis et al. 2011) | $MR1^+$ bacterial small molecules (Kjer-Nielsen et al. 2012), *M. tuberculosis* (Gold and Lewinsohn 2011) |

(continued)

**Table 1** (continued)

| Subset name | Definition | Surface marker expression | Functional capacities | Trafficking properties | Example antigen specificities |
|---|---|---|---|---|---|
| iNKT cells | A subset of NKT cells that express a highly-restricted TCR and respond to self or foreign CD1d-restricted lipid ligand (Juno et al. 2012; Brennan et al. 2013) | $CD161^+$, Va24Ja18 (Juno et al. 2012) $CD4^+$, $CD8^-$ or DN T cells (Juno et al. 2012), Memory/effector phenotype (Brennan et al. 2013) | Respond in an innate-like manner to danger signals and proinflammatory cytokines and exert their effector functions within hours of being activated (Brennan et al. 2013) | Preferentially localize to non-lymphoid tissues (Gold and Lewinsohn 2011), $CD4^-$ subset are $CD62L^{lo}$ with higher levels of CD11a (tissue-infiltrating phenotype); $CD4^+$ subset are $CD62L^{hi}$ (lymph node homing phenotype) (Juno et al. 2012) | CD1d and modified lipids, depletion in HIV and HCV (Juno et al. 2012) |

Without strict standards and complete understanding of what functionally distinguishes each of these subsets, it is inevitable that certain subsets will be redundant. The process of distinguishing these cell subsets and filling in blank values in this table will be facilitated by high-dimensional analysis techniques such as mass cytometry. We favor a continuum model whereby the distributions of phenotypic and functional properties of each of these subsets are often overlapping. Excluding invariant T cells, we arranged the cell subsets in an order that approximates their interrelatedness is supported by differential phenotypic compositions and contributions to primary versus secondary responses (Buchholz et al. 2013; Gerlach et al. 2013). Linkages between cellular phenotypes and cell fates of activated T cells are more difficult to study in humans than in mouse models and may vary depending on the nature of the infection. We think that the measurement of many more properties of responding T cells will facilitate incorporation of concepts derived from mouse studies in a way that is applicable to clinical use.

Considerable effort has also been dedicated to modes and mechanisms of T cell tolerance and dysfunction. This research has clear ramifications for treatment of graft rejection, chronic infection, and cancer immune evasion. Focusing on $CD8^+$ T cell dysfunction, several terms have arisen including anergy (Schwartz 2003), exhaustion (Wherry 2011), and senescence (Nikolich-Zugich et al. 2012; Goronzy and Weyand 2013), each with different but often overlapping definitions (Akbar and Henson 2011; Crespo et al. 2013b) (Table 1). Here again, our hope is that an ability to simultaneously assess a large number of markers associated with anergy, exhaustion, and/or senescence in conjunction with single-cell measures of functional attributes will allow for better delineation of T cell fates. Such an approach requires the study of a range of well-defined immune responses and an ability to probe a broad range of phenotypic markers to cover the elusive hallmarks of each state, which is a significant challenge.

T cells can also be delineated by their tissue trafficking capacities. In human immunological studies, by far the most common tissue available for interrogation is blood. Fortunately, T cells traffic through the blood during various phases of immune surveillance and response, providing an opportunity to monitor cells participating in each of these phases. As described above, current paradigms used for defining the status of blood T cells rely on lymph node trafficking receptors CCR7 and CD62L (Sallusto et al. 1999; Masopust and Schenkel 2013). One can also delineate cells trafficking to other tissues (or locations within tissues) by analysis of expression of other trafficking receptors. These receptors, which provide specificity to tissue entry through blood vessel walls, are involved in tethering (integrins or selectins), reception of signal from tissue-derived chemokines (chemokine receptors), and adhesion (integrins), and allow for arrest and diapedesis in a tissue-specific manner. Specificity of this process may be derived from the combinatorial diversity of trafficking receptor expression on T cells (Butcher and Picker 1996). The trafficking receptor expression profiles are programmed during initial T cell priming. For instance, T cell priming in the presence of retinoic acid, a small intestine-derived vitamin A derivative, results in T cell small intestinal homing through expression of the trafficking receptors integrin $\alpha 4\beta 7$ and chemokine receptor CCR9. Similarly, vitamin D metabolites, dependent on ultraviolet light exposure for formation and present at high levels in the skin, are involved in imprinting T cells with skin epidermal homing capacity through elevated expression of CCR10, which cooperates with CLA and CCR4 (also expressed on skin-homing T cells) (Sigmundsdottir and Butcher 2008). Given the range of tissues and locations within tissues that are under surveillance by T cells, regulation of T cell trafficking is certainly a complicated task. High-dimensional analysis using mass cytometry will allow simultaneous assessment of many trafficking receptors in T cells, leading to a more thorough analysis of the complicated process of T cell trafficking.

## 3 High-Dimensional Analysis of T Cell Phenotype and Function

As described above, the use of high-dimensional analysis techniques applied to the study of T cells promises to further our understanding of the overwhelming diversity of this cell type. Until the throughput of single-cell gene expression profiling improves and single-cell mass spectrometry-based peptide-sequencing proteomics are optimized (Dominguez et al. 2013), flow cytometry is the most effective method for assessing the diversity of T cells. An ability to probe up to $\sim$18 cellular parameters at >10,000 cells per second has allowed simultaneous assessment of T cell surface markers and intracellular markers including transcription factors and cytokines (Chattopadhyay and Roederer 2012). This has greatly facilitated studies on the relationships between T cell phenotype/function and clinical status for a range of diseases. For instance, to identify a correlation between the capacity of an antigen-specific T cell to produce particular cytokines and the extent of disease progression or protective immunity requires the ability to probe at least nine cellular parameters (Betts et al. 2006; Seder et al. 2008; Yuan et al. 2008; Makedonas and Betts 2011). With the 12–18 parameters (dimensions) available using polychromatic flow cytometry, we are well into a range of dimensionality that is incomprehensible without the assistance of computer-aided analysis. There is currently no "right way" to analyze the high-dimensional data generated, and efforts are underway to develop tools to visualize and understand the data and to deal with the onslaught of flow cytometry data being generated (Aghaeepour et al. 2013).

Enter mass cytometry: a mass spectrometry-based flow cytometry method that uses isotopically purified heavy metal atoms as tags and allows analysis of more than 30 cellular markers per cell and promises up to 100. The technical benefits of mass cytometry are clear, starting with the greatly reduced cross talk between channels, which in turn simplifies analysis and interpretation of the data. With so many colors (parameters) in fluorescence flow cytometry and dimensionality that is already incomprehensible, why do we need to extend this even further? To answer this question, we must consider again the categories of T cell diversity described above: T cell specificity (10,000–1 million possible), antigen experience, memory versus. effector fate, functional capacity, extent of dysfunction (i.e., anergy, exhaustion, senescence), and tissue-specific trafficking. Immediately, it is clear that even 30 or more markers are insufficient to probe all of the important aspects of T cell biology. Thus, we must choose based on the application of the day.

As a start, mass cytometry has been adapted to probe the relationship between T cell phenotype and function through the simultaneous assessment of many markers of each. After several parameters were used to isolate single CD8$^+$ T cells, this population was probed with six different peptide-MHC tetramers (discussed below), and 25 parameters were used to evaluate the cells (Newell et al. 2012). These 25 parameters included 16 surface markers variably expressed on human CD8$^+$ T cells and nine different functional markers. Functional indicators included intracellular staining for expression of cytotoxic granule components (such as granzymes and perforin) (Peters et al. 1991), transiently exposed intravesicular CD107 molecules to measure the secretion of the lytic granules upon stimulation (Betts et al. 2003), and intracellular staining for expression of stimulation-induced cytokine (IL-2, GM-CSF, MIP-1$\alpha$, MIP-1$\beta$, TNF-$\alpha$, and IFN-$\gamma$) production (Waldrop et al. 1997; De Rosa et al. 2004; Betts et al. 2006). Extending the simultaneous analysis of T cell multifunctionality to these nine different functional capacities, there are $2^9$, for a total of 512, different possible combinations of function that can be produced by a given cell. Depending on the cellular surface phenotype, type of stimulation, or the antigen specificity of the cells, the combinatorial repertoire of functional capacities is skewed for each population of cells (Newell et al. 2012).

Pursuing an integrated analysis of all phenotypic and functional capacities of the cells, all data were aggregated and subjected to dimensionality reduction using PCA. Through linear combination of all parameters measured for each cell, new composite measures were created. The result is that information content in the data set can be maximally represented by a minimal number of dimensions (principal components). Although simple and effective, we anticipate that better resolution of cellular subsets will be obtained by the use of methods that better account for nonlinear relationships between cellular markers (Amir et al. 2013). Nonetheless, by this analysis, a consistent overall pattern was observed in each of the donors tested. By looking at where previously characterized populations of CD8$^+$ T cells were located within the pattern, the meaning for this pattern was apparent. In short, two major discrete populations were identified: a homogenous population of naïve T cells and a highly heterogeneous population of memory/effector cells. Within the memory/effector cell population, a continuous distribution of cells were observed corresponding to a phenotypic and functional progression. At one extreme were IL-2-producing cells with central memory-like phenotypes and on the other were cells with phenotypes similar to previously described short-lived effector cells. By plotting average expression of various phenotypic markers and functional capacities along the continuum of memory/effector cells, a progression for each marker was described that was used to define a hypothesis of how memory/effector T cells are related to each other (Fig. 1).

By overlaying various antigen-specific cells (as discussed in Sect. 4) over these PCA plots, the general phenotypic and functional properties of the antigen-specific cells could be objectively compared. This analysis provided additional validation for the approach by showing that cells specific for influenza, a sporadic infection, were on one end of the memory/effector cell spectrum, whereas cells specific for chronic viral infections like Epstein-Barr virus (EBV) and cytomegalovirus

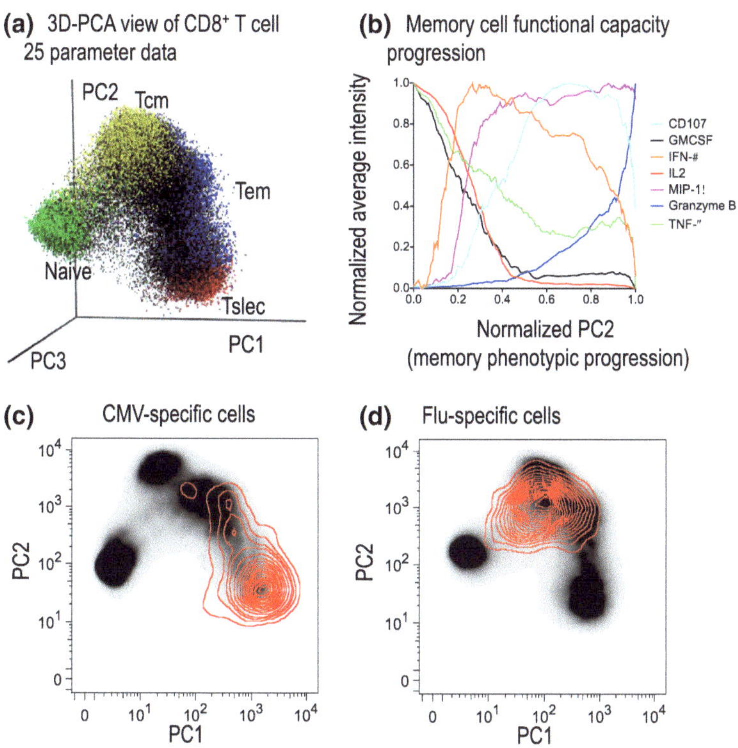


**Fig. 1** Phenotypic and functional diversity of total memory and antigen-specific CD8+ T cells shown as three-dimensional PCA mass cytometry data. Peripheral blood mononuclear cells from healthy donors were stimulated with PMA and ionomycin and stained for 25 phenotypic and functional markers. PCA explained approximately 60 % of the variance. **a** The PCA was performed and the values for the first three components are plotted for each cell. Cells were gated by surface marker phenotype. Naïve (*green*), central memory (*yellow*), effector memory (*blue*), and terminally differentiated effector (*red*) cell populations are overlaid to identify the main phenotypic clusters. **b** To analyze only memory cells, cells in the principal components PC1 versus PC2 plot in *panel A* were gated to exclude the naive compartment (cells with low value for PC1). The average expression level of several functional markers are plotted versus PC2, illustrating a functional progression within the CD8+ T cell compartment. **c**, **d** Density plots of the first two PCA components of total live CD8+ T cells are overlaid with contour density plots of peptide-MHC tetramer-positive cells (*red*). **c** CMV-specific cells. **d** Influenza-specific cells. (Newell et al. 2012)

(CMV) were distributed along the progression depending on the donor and the type of infection (Newell et al. 2012). The similarities of overall phenotypic patterns of additional T cell specificities from additional normal donors shows that PCA analysis can also discriminate EBV-specific cells targeting latency-associated versus lytic cycle-derived antigens (Newell et al. 2013) in agreement with a previous analysis of T cell responses during acute EBV infection (Hislop et al. 2002). With the exception of Mart-1-specific cells (Newell et al. 2013), very few antigen-specific cells occupy the naïve compartment of the 3D-PCA space, which is expected (Newell et al. 2012). That influenza-, CMV- or EBV-specific cells were detected suggest that these virus-specific cells exhibit a phenotypic reversion. An additional population of cells near the central memory portion of the memory/effector population is also observed to be lacking in cells specific for any antigen probed (Newell et al. 2012). This cluster of $CD49d^{-}CD45RA^{-}CD45RO^{+}$-$CD28^{+}IL\text{-}2^{-}$ cells may correspond to MAIT cells, which represent a significant portion of blood $CD8^{+}$ T cells. MAIT cells express semi-invariant TCRs and should not be stained by any of the peptide-MHC tetramers that were used in the study (Gold and Lewinsohn 2013). In our study, the omission of an anti-CD161 and anti-$V\alpha7.2$ antibody prevented a conclusive identification of this cluster of cells. However, this example is an indication of the power of the high-dimensional approach to identify populations by combinations of markers even when certain useful markers are left out of the analysis panel.

# 4 Identifying and Characterizing Antigen-Specific T Cells

T cell antigen specificity is conferred by the TCR for the peptide-MHC complex. In order for the antigen to be recognized by the TCR, it must be broken down into various peptides and then bound to the MHC. Accurate a priori prediction of the identities of antigens (i.e., peptide epitopes) targeted by T cells during an immune response is difficult, if not impossible. The difficulties in prediction of peptide epitopes are due to factors such as age, viral mutation, and route of infection that influence the repertoire of epitopes presented to T cells and the availability of T cells to respond to the epitopes (Chang et al. 2011; Akram and Inman 2012). Knowledge of antigenic targets of the T cell-mediated immune response confers an ability to track, understand, and identify ways to manipulate the antigen-specific immune response (Davis et al. 2011). For instance, the identification of common antigen specificities and phenotypic and functional features of the T cell-mediated immune responses that correlate with disease outcomes could be useful as early diagnostic indicators. This may be especially applicable to early diagnosis of cancer or for evaluation of cancer immunotherapy efficacy (Sharma et al. 2011; Heemskerk et al. 2013). Knowledge of the T cell targets of inflammatory diseases could lead to novel therapies aimed at interfering with target recognition (Monteleone et al. 2011). Unfortunately, very few disease-associated T cell epitopes have been unambiguously identified. Here, we discuss recent advances in methods of identifying antigen-specific T cells as a means to exploit these cells for both basic and clinical immunology purposes.

Traditional means of identifying antigen-specific T cells are indirect and rely on a cellular response to stimulation. These antigen-specific responses include: proliferation, cytokine production, target cell lysis (killing), or changes in surface marker expression (e.g., CD107, CD40L, CD69, CD137) (Brunner et al. 1968; Peters et al. 1991; Waldrop et al. 1997; Betts et al. 2003; De Rosa et al. 2004; Frentsch et al. 2005; Betts et al. 2006). One major technical advantage of using cellular responses as a means of identifying antigen-specific cells is that neither the precise epitope nor the restriction element needs to be known. Then, by an iterative process these

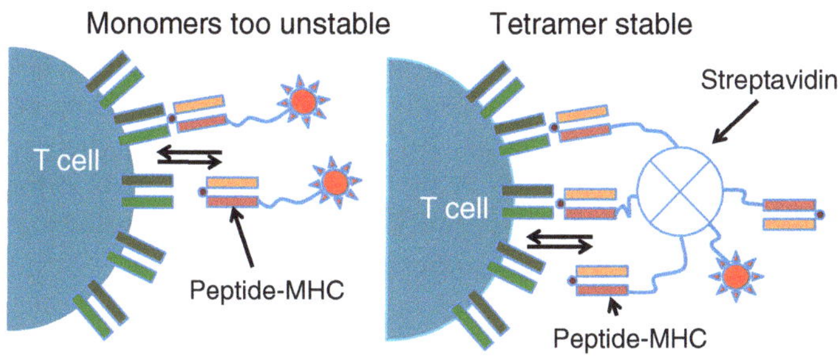


**Fig. 2** Peptide MHC-tetramer staining concept. The affinity of monomeric peptide-MHC for TCR (expressed on T cells) is low, usually between 5 and 100 μM. Cooperative binding overcomes this limitation and allows for robust direct detection of antigen-specific cells

unknowns can be determined. Disadvantages are numerous: (1) Such an iterative process of epitope mapping can be tedious and requires a large amount of cellular material. (2) The sensitivity (i.e., frequency of antigen-specific cells that can be detected) varies with assay type but depends heavily on the extent of background T cell activation. (3) Only cells with a given detectable response can be identified, precluding detection of dysfunctional cells. (4) Unless cells are sorted and subjected to further analysis, the phenotypic and/or functional profile of the antigen-specific cells are usually very limited. (5) The stimulation used to identify the antigen-specific cell can itself alter the phenotypic profile of the cells being studied.

Many of these limitations have been overcome by improvements to methods of direct detection of antigen-specific cells with peptide-MHC multimers. Although T cells are exquisitely sensitive, the range of affinities that TCRs have for cognate peptide MHC are remarkably low (5–100 μM). Furthermore, because the off-rates of TCR-peptide MHC interaction are often very fast (time constants usually only a few seconds) labeling antigen-specific T cells with monomeric peptide-MHC complexes proved to be impossible. However, when peptide-MHC molecules are multimerized, cooperative binding allows for robust and specific labeling of antigen-specific T cells (Fig. 2). This can be achieved by the use of biotinylated peptide-MHC complexes tetramerized with fluorescently labeled streptavidin molecules (peptide-MHC tetramers), which allow direct identification of unperturbed antigen-specific T cells by flow cytometry (Altman et al. 1996). The application of peptide-MHC tetramer or multimer staining has enabled a number of labs to reliably monitor the status of T cells for a variety of immune responses (Lee et al. 1999; Clay et al. 2001; Hobeika et al. 2001; Appay et al. 2008; Bolton and Roederer 2009).

A number of recent advances have overcome several long-standing obstacles in peptide-MHC tetramer staining technology. The development of UV-sensitive, exchangeable peptides greatly simplifies the production of hundreds or thousands of tetramers from one batch of prepared MHC Class I protein (Toebes et al. 2006; Grotenbreg et al. 2008). For detection of rare antigen-specific T cell populations, a magnet-based enrichment procedure has allowed the characterization of very rare

T cells, even those in the naïve repertoire (Day et al. 2003; Moon et al. 2007). To probe a larger number of T cell antigen specificities in a single sample, combinatorial staining approaches have recently been developed (Hadrup et al. 2009; Newell et al. 2009; Andersen et al. 2012) and improved by the use of mass cytometry (Newell et al. 2013). With the combined use of mass cytometry and peptide-MHC multimers, antigen-specific T cells can be interrogated together with their phenotypic and functional properties.

## 5 Multiplexed Assessment of T Cell Antigen Specificity

In addition to an ability to probe many more markers on each cell, a major strength of mass cytometry is the ability to multiplex by using combinations of tags as specific markers. This strength has been remarkably demonstrated with mass-tag barcoding, allowing for high throughput analysis of enumerable cell types subjected to a vast array of cellular stimulation conditions analyzed in parallel (Bodenmiller et al. 2012). In another form of multiplexing, the utility of mass cytometry for greatly extending combinatorial tetramer staining was recently demonstrated. By dedicating ten of the ~39 useful heavy-isotope channels available to combinatorially encoded peptide-MHC tetramers, more than 100 T cell antigen specificities were probed in blood or intestinal lymphocyte samples (Fig. 3). With the use of magnetic cellular enrichment, antigen-specific cells at a frequency as low as one in 100,000 $CD8^+$ T cells were detected and characterized (Newell et al. 2013). By this approach, T cells specific for rotavirus epitopes present in the blood of healthy donors and intestinal biopsies of 21 patients samples were identified and interrogated with more than 20 different phenotypic markers.

In addition to the capacity to screen for new T cell epitopes, mass cytometry-based high content analysis of antigen-specific cells benefits from the ability to simultaneously probe the phenotypic profiles of a large number of previously defined T cell epitopes. As described in Sect. 3 of this chapter, the relationship between phenotypic marker expression and the actual functional status of T cells is not always immediately apparent. When characterizing antigens such as influenza, CMV, EBV, or Mart-1, the status of infection for normal donors are usually known and can be used as landmarks to define the phenotypic profiles observed for these cells. For instance, almost all normal donors have been exposed to and have cleared the influenza virus at least once. The phenotypes of influenza-specific cells from donors who have not been exposed to influenza recently are similar and match the expected phenotype of central memory cells ($CD45RA^-CCR7^+$). Similar logic can be used to decipher the meaning of the phenotypes of CMV- and EBV-specific cells, both represent cells indicative of latent chronic viral infections. Mart-1-specific cells are well-studied T cells with an unusually high naïve precursor frequency (Cole et al. 2009; Alanio et al. 2010). An additional benefit for probing a number of previously defined T cell epitopes is that it serves as an internal control for antigen-independent effects. For instance, bystander activation

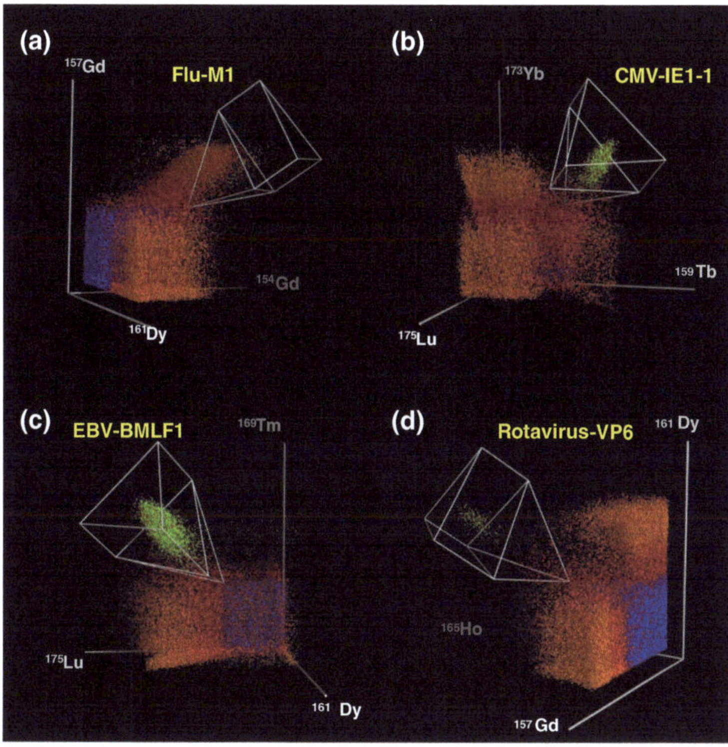


**Fig. 3** Three dimensional combinatorial tetramer staining example plots. **a** Flu-M1, **b** CMV-IE1, **c** EBV-BMLF1, and **d** rotavirus-VP6 specific cells were detected in a sample from one donor. Each T cell antigen specificity is tagged with peptide-MHC tetramers made with a unique combination of three of 10 possible metal-conjugated streptavidin molecules. Flu-M1 specific cells are triple-stained with tetramers loaded with Sm-154, Gd-157, and Dy-161 metals (similar codes for the other epitopes are as follows: CMV-IE1: Tb-159, Lu-175, Yb-173. EBV: Dy-161, Tm-169, Lu-175. Rotavirus-VP6: Gd-157, Dy-161, Ho-165). Cells positive for any other metal are colored *red* and excluded. Cells that are negative for all 10 metal labels are colored *blue*, cells colored *orange* were identified as specific for a different epitope, and cells that meet all criteria are colored *green*. The *white box* is an approximate 3D representation of the gate used to delineate these cells. (Newell et al. 2013)

has been observed during acute HIV and other infections (Doisne et al. 2004; Sandalova et al. 2010).

As demonstrated for human rotavirus, mass cytometry-based combinatorial tetramer staining is a useful method for T cell epitope screening (Newell et al. 2013). With traditional methods of epitope screening, additional blood samples are needed for validation of hits and for follow-up analysis of T cell phenotypes. Using mass cytometry-based combinatorial tetramer staining, hits can be validated in several ways. One way is to split each sample and use an alternate coding scheme for each epitope. Hits with good correspondence between the two coding schemes can be counted as true positives. This method controls for several sources of artifact. Another way to validate hits is to evaluate the phenotypes of the cells identified. In the case of rotavirus, intestinal trafficking receptor co-expression served as a method for validating many of the hits.

So far, 10 of the ~39 or more heavy metal channels available for mass cytometry have been dedicated to analysis of antigen-specific cells with tetramers. This approach allows for analysis of 120 specificities if each T cell epitope is tagged with a unique combination of three tags: given 10 tags, there are 120 ways of creating unique sets of three. There are no obvious reasons why this concept could not be extended to greatly increase the number of specificities that could be probed in the same sample. The number may be limited by loss of signal with ever-increasing concentrations of irrelevant tetramer. To test this, we performed a titration that showed sufficient signal even when 1,000-fold excess of irrelevant tetramer was present. Additional combinatorial capacity can also be achieved by coding each antigen specificity with four or more metal tags. By this approach, the overall tetramer signal will be diluted due to competition for TCR binding on the cell. However, as described (Newell et al. 2013), using additional metals for each antigen specificity has the advantage of reducing background. That is because the background noise for each channel is mostly random and events with coincident signal in four or more channels are extremely rare except in cases of true positivity. We suggest that it should soon be feasible to probe for over 1,000 T cell specificities by coding each antigen with a unique combination of four metals chosen from a list of 15 (15 tags allow 1,365 unique sets of 4), and this may not even be the ultimate limit to this approach. Dedicating more dimensional space than needed for the number of specificities being probed may also be useful for increasing statistical power of hit validation as described above.

# 6 Broad Applicability of a Combined Approach and Future Perspectives

As emphasized throughout this chapter, the T cell lineage is arguably one of the most diverse cell types in the body. Making sense out of this large T cell diversity for meaningful and useful categorizations is indeed challenging. However, it is also critical for understanding the functional role of T cells in immune responses and for better assessment of the status of ongoing T cell-mediated immune responses. Mass cytometry brings an ability to measure an unprecedented number of cellular parameters at high cellular throughput and is well suited for the delineation of various T cell subsets. Here, we also described the combinatorial tetramer staining approach, which also takes advantage of the multiplexing capacity of the mass cytometry platform. By combining an ability to deeply probe the phenotypes, trafficking receptor profiles, functional capacities, and antigen specificities of T cells, all of the most important features of the T cell response can be simultaneously determined including: How recently has the cell been exposed to antigen stimulation? How much has the T cell differentiated? To what tissue is the T cell migrating? What type of functional response is the T cell mounting? What is the T cell's precise specificity? The answers to these questions will provide fundamental insight to the status of any T cell-mediated immune response as well as an indication of the type of contribution the T cells are making to the response. Although this chapter has been focused on CD8$^+$ T cell response, mass cytometry-based high-dimensional analysis as described here can be applied to any other type of T cell. For less conventional T cell subsets, less phenotypic and functional diversity may be expected. In contrast, for the CD4$^+$ T cell response, it will be especially interesting to see how high-dimensional analysis can address the interrelatedness of the growing number helper T cell subsets and extreme potential for diversity.

Going forward, we expect that technical improvements will allow many more cellular parameters to be probed simultaneously. Mass cytometry offers many atomic mass channels that will soon be filled with relevant cellular probes. Advancement will also come in our ability to more accurately detect less abundant cellular proteins, and this will allow a broader range of markers to be probed in T cells. For instance, accurate assessment of many more trafficking receptors, transcription factors, low abundance receptors, and intracellular signaling molecules should become easier to analyze. For T cell epitope discovery, we think this approach will nicely complement proteomics-based approaches for identification of MHC-binding candidate T cell epitopes (Hillen and Stevanovic 2006; Hoppes et al. 2010). Unlike epitope prediction algorithms, elution and direct identification of endogenously MHC-bound peptides will provide an accurate means of identifying candidate T cell epitopes, whose roles in the immune response can be directly tested using combinatorial tetramer staining as described here. In sum, we should soon be able to detect many more antigen specificities and probe each of the cells with many more phenotypic, functional, and signaling molecules allowing a fully integrated perspective of the diversity of the T cell compartment with a better ability to resolve closely related T cell subsets present in states of health and disease.

## References


Agace WW (2006) Tissue-tropic effector T cells: generation and targeting opportunities. Nat Rev Immunol 6:682–692

Aghaeepour N, Finak G, Hoos H, Mosmann TR, Brinkman R, Gottardo R, Scheuermann RH (2013) Critical assessment of automated flow cytometry data analysis techniques. Nat Methods 10:228–238

Akbar AN, Henson SM (2011) Are senescence and exhaustion intertwined or unrelated processes that compromise immunity? Nat Rev Immunol 11:289–295

Akbar AN, Terry L, Timms A, Beverley PC, Janossy G (1988) Loss of CD45R and gain of UCHL1 reactivity is a feature of primed T cells. J Immunol 140:2171–2178

Akram A, Inman RD (2012) Immunodominance: a pivotal principle in host response to viral infections. Clin Immunol 143:99–115

Alanio C, Lemaitre F, Law HK, Hasan M, Albert ML (2010) Enumeration of human antigen-specific naive CD8+ T cells reveals conserved precursor frequencies. Blood 115:3718–3725

Altman JD, Moss PA, Goulder PJ et al (1996) Phenotypic analysis of antigen-specific T lymphocytes. Science 274:94–96

Amir EA, Davis KL, Tadmor MD et al (2013) viSNE enables visualization of high dimensional single-cell data and reveals phenotypic heterogeneity of leukemia. Nat Biotechnol 31:545–552

Andersen RS, Kvistborg P, Frosig TM et al (2012) Parallel detection of antigen-specific T cell responses by combinatorial encoding of MHC multimers. Nat Protoc 7:891–902

Anichini A, Molla A, Vegetti C et al (2010) Tumor-reactive CD8+ early effector T cells identified at tumor site in primary and metastatic melanoma. Cancer Res 70:8378–8387

Anichini A, Scarito A, Molla A, Parmiani G, Mortarini R (2003) Differentiation of CD8+ T cells from tumor-invaded and tumor-free lymph nodes of melanoma patients: role of common gamma-chain cytokines. J Immunol 171:2134–2141

Appay V, Douek DC, Price DA (2008) CD8+ T cell efficacy in vaccination and disease. Nat Med 14:623–628

Appay V, Dunbar PR, Callan M et al (2002) Memory CD8+ T cells vary in differentiation phenotype in different persistent virus infections. Nat Med 8:379–385

Appay V, Nixon DF, Donahoe SM et al (2000) HIV-specific CD8(+) T cells produce antiviral cytokines but are impaired in cytolytic function. J Exp Med 192:63–75

Barber DL, Wherry EJ, Masopust D et al (2006) Restoring function in exhausted CD8 T cells during chronic viral infection. Nature 439:682–687

Bengsch B, Seigel B, Ruhl M et al (2010) Coexpression of PD-1, 2B4, CD160 and KLRG1 on exhausted HCV-specific CD8+ T cells is linked to antigen recognition and T cell differentiation. PLoS Pathog 6:e1000947

Betts MR, Brenchley JM, Price DA, De Rosa SC, Douek DC, Roederer M, Koup RA (2003) Sensitive and viable identification of antigen-specific CD8+ T cells by a flow cytometric assay for degranulation. J Immunol Methods 281:65–78

Betts MR, Nason MC, West SM et al (2006) HIV nonprogressors preferentially maintain highly functional HIV-specific CD8+ T cells. Blood 107:4781–4789

Blackburn SD, Shin H, Haining WN et al (2009) Coregulation of CD8+ T cell exhaustion by multiple inhibitory receptors during chronic viral infection. Nat Immunol 10:29–37

Bodenmiller B, Zunder ER, Finck R et al (2012) Multiplexed mass cytometry profiling of cellular states perturbed by small-molecule regulators. Nat Biotechnol

Bolton DL, Roederer M (2009) Flow cytometry and the future of vaccine development. Expert Rev Vaccin 8:779–789

Brenchley JM, Karandikar NJ, Betts MR et al (2003) Expression of CD57 defines replicative senescence and antigen-induced apoptotic death of CD8+ T cells. Blood 101:2711–2720

Brennan PJ, Brigl M, Brenner MB (2013) Invariant natural killer T cells: an innate activation scheme linked to diverse effector functions. Nat Rev Immunol 13:101–117

Brunner KT, Mauel J, Cerottini JC, Chapuis B (1968) Quantitative assay of the lytic action of immune lymphoid cells on 51-Cr-labelled allogeneic target cells in vitro; inhibition by isoantibody and by drugs. Immunology 14:181–196

Buchholz VR, Flossdorf M, Hensel I et al (2013) Disparate individual fates compose robust CD8+ T cell immunity. Science

Butcher EC, Picker LJ (1996) Lymphocyte homing and homeostasis. Science 272:60–66

Chang CX, Dai L, Tan ZW, Choo JA, Bertoletti A, Grotenbreg GM (2011) Sources of diversity in T cell epitope discovery. Front Biosci 16:3014–3035 A journal and virtual library

Chattopadhyay PK, Roederer M (2012) Cytometry: today's technology and tomorrow's horizons. Methods 57:251–258

Clay TM, Hobeika AC, Mosca PJ, Lyerly HK, Morse MA (2001) Assays for monitoring cellular immune responses to active immunotherapy of cancer. Clini Cancer Res 7:1127–1135 An Official Journal of the American Association for Cancer Research

Cole DK, Yuan F, Rizkallah PJ et al (2009) Germ line-governed recognition of a cancer epitope by an immunodominant human T-cell receptor. J Biol Chem 284:27281–27289

Crespo J, Sun H, Welling TH, Tian Z, Zou W (2013a) T cell anergy, exhaustion, senescence, and stemness in the tumor microenvironment. Curr Opin Immunol 25:214–221

Crespo J, Sun H, Welling TH, Tian Z, Zou W (2013b) T cell anergy, exhaustion, senescence, and stemness in the tumor microenvironment. Curr Opin Immunol
Davis MM, Altman JD, Newell EW (2011) Interrogating the repertoire: broadening the scope of peptide-MHC multimer analysis. Nat Rev Immunol 11:551–558
Davis MM, Bjorkman PJ (1988) T-cell antigen receptor genes and T-cell recognition. Nature 334:395–402
Day CL, Seth NP, Lucas M et al (2003) Ex vivo analysis of human memory CD4 T cells specific for hepatitis C virus using MHC class II tetramers. J Clin Invest 112:831–842
De Rosa SC, Lu FX, Yu J et al (2004) Vaccination in humans generates broad T cell cytokine responses. J Immunol 173:5372–5380
Derhovanessian E, Maier AB, Hahnel K et al (2011) Infection with cytomegalovirus but not herpes simplex virus induces the accumulation of late-differentiated CD4$^+$ and CD8+ T-cells in humans. J Gen Virol 92:2746–2756
Derre L, Rivals JP, Jandus C et al (2010) BTLA mediates inhibition of human tumor-specific CD8+ T cells that can be partially reversed by vaccination. J Clin Investig 120:157–167
Doisne JM, Urrutia A, Lacabaratz-Porret C et al (2004) CD8+ T cells specific for EBV, cytomegalovirus, and influenza virus are activated during primary HIV infection. J Immunol 173:2410–2418
Dominguez MH, Chattopadhyay PK, Ma S et al (2013) Highly multiplexed quantitation of gene expression on single cells. J Immunol Methods 391:133–145
Ely KH, Cookenham T, Roberts AD, Woodland DL (2006) Memory T cell populations in the lung airways are maintained by continual recruitment. J Immunol 176:537–543
Feng D, Bond CJ, Ely LK, Maynard J, Garcia KC (2007) Structural evidence for a germline-encoded T cell receptor-major histocompatibility complex interaction ‘codon’. Nat Immunol 8:975-983
Frentsch M, Arbach O, Kirchhoff D et al (2005) Direct access to CD4$^+$ T cells specific for defined antigens according to CD154 expression. Nat Med 11:1118–1124
Garcia KC, Adams JJ, Feng D, Ely LK (2009) The molecular basis of TCR germline bias for MHC is surprisingly simple. Nat Immunol 10:143–147
Gattinoni L, Lugli E, Ji Y et al (2011) A human memory T cell subset with stem cell-like properties. Nat Med 17:1290–1297
Gebhardt T, Wakim LM, Eidsmo L, Reading PC, Heath WR, Carbone FR (2009) Memory T cells in nonlymphoid tissue that provide enhanced local immunity during infection with herpes simplex virus. Nat Immunol 10:524–530
Gerlach C, Rohr JC, Perie L et al (2013) Heterogeneous differentiation patterns of individual CD8+ T cells. Science
Gillespie GM, Wills MR, Appay V et al (2000) Functional heterogeneity and high frequencies of cytomegalovirus-specific CD8(+) T lymphocytes in healthy seropositive donors. J Virol 74:8140–8150
Gold MC, Lewinsohn DM (2011) Mucosal associated invariant T cells and the immune response to infection. Microbes Infect Inst Pasteur 13:742–748
Gold MC, Lewinsohn DM (2013) Co-dependents: MR1-restricted MAIT cells and their antimicrobial function. Nat Rev Microbiol 11:14–19
Golden-Mason L, Palmer BE, Kassam N et al (2009) Negative immune regulator Tim-3 is overexpressed on T cells in hepatitis C virus infection and its blockade rescues dysfunctional CD4 + and CD8+ T cells. J Virol 83:9122–9130
Goronzy JJ, Weyand CM (2013) Understanding immunosenescence to improve responses to vaccines. Nat Immunol 14:428–436
Grotenbreg GM, Roan NR, Guillen E et al (2008) Discovery of CD8+ T cell epitopes in Chlamydia trachomatis infection through use of caged class I MHC tetramers. Proc Natl Acad Sci U S A 105:3831–3836
Hadrup SR, Bakker AH, Shu CJ et al (2009) Parallel detection of antigen-specific T-cell responses by multidimensional encoding of MHC multimers. Nat Methods 6:520–526

Hamann D, Baars PA, Rep MH, Hooibrink B, Kerkhof-Garde SR, Klein MR, van Lier RA (1997) Phenotypic and functional separation of memory and effector human CD8+ T cells. J Exp Med 186:1407–1418
Han A, Newell EW, Glanville J, Fernandez-Becker N, Khosla C, Chien YH, Davis MM (2013) Dietary gluten triggers concomitant activation of CD4 + and CD8+ alphabeta T cells and gammadelta T cells in celiac disease. In: Proceedings of the national academy of sciences of the United States of America
Heemskerk B, Kvistborg P, Schumacher TN (2013) The cancer antigenome. EMBO J 32:194–203
Heffner M, Fearon DT (2007) Loss of T cell receptor-induced Bmi-1 in the KLRG1(+) senescent CD8(+) T lymphocyte. Proc Natl Acad Sci USA 104:13414–13419
Henson SM, Riddell NE, Akbar AN (2012) Properties of end-stage human T cells defined by CD45RA re-expression. Curr Opin Immunol 24:476–481
Hillen N, Stevanovic S (2006) Contribution of mass spectrometry-based proteomics to immunology. Expert Rev Proteomics 3:653–664
Hislop AD, Annels NE, Gudgeon NH, Leese AM, Rickinson AB (2002) Epitope-specific evolution of human CD8(+) T cell responses from primary to persistent phases of Epstein-Barr virus infection. J Exp Med 195:893–905
Hobeika AC, Clay TM, Mosca PJ, Lyerly HK, Morse MA (2001) Quantitating therapeutically relevant T-cell responses to cancer vaccines. Crit Rev Immunol 21:287–297
Holler PD, Kranz DM (2004) T cell receptors: affinities, cross-reactivities, and a conformer model. Mol Immunol 40:1027–1031
Hoppes R, Ekkebus R, Schumacher TN, Ovaa H (2010) Technologies for MHC class I immunoproteomics. J Proteomics 73:1945–1953
Jones RB, Ndhlovu LC, Barbour JD et al (2008) Tim-3 expression defines a novel population of dysfunctional T cells with highly elevated frequencies in progressive HIV-1 infection. J Exp Med 205:2763–2779
Joshi NS, Cui W, Chandele A et al (2007) Inflammation directs memory precursor and short-lived effector CD8(+) T cell fates via the graded expression of T-bet transcription factor. Immunity 27:281–295
Juno JA, Keynan Y, Fowke KR (2012) Invariant NKT cells: regulation and function during viral infection. PLoS Pathog 8:e1002838
Kaech SM, Cui W (2012) Transcriptional control of effector and memory CD8+ T cell differentiation. Nat Rev Immunol 12:749–761
Kaech SM, Hemby S, Kersh E, Ahmed R (2002) Molecular and functional profiling of memory CD8 T cell differentiation. Cell 111:837–851
Kaech SM, Tan JT, Wherry EJ, Konieczny BT, Surh CD, Ahmed R (2003) Selective expression of the interleukin 7 receptor identifies effector CD8 T cells that give rise to long-lived memory cells. Nat Immunol 4:1191–1198
Kjer-Nielsen L, Patel O, Corbett AJ et al (2012) MR1 presents microbial vitamin B metabolites to MAIT cells. Nature 491:717–723
Le Bourhis L, Guerri L, Dusseaux M, Martin E, Soudais C, Lantz O (2011) Mucosal-associated invariant T cells: unconventional development and function. Trends Immunol 32:212–218
Le Bourhis L, Mburu YK, Lantz O (2013) MAIT cells, surveyors of a new class of antigen: development and functions. Curr Opin Immunol 25:174–180
Lee PP, Yee C, Savage PA et al (1999) Characterization of circulating T cells specific for tumor-associated antigens in melanoma patients. Nat Med 5:677–685
Li H, Wu K, Tao K et al (2012) Tim-3/galectin-9 signaling pathway mediates T-cell dysfunction and predicts poor prognosis in patients with hepatitis B virus-associated hepatocellular carcinoma. Hepatology 56:1342–1351
Makedonas G, Betts MR (2011) Living in a house of cards: re-evaluating CD8+ T-cell immune correlates against HIV. Immunol Rev 239:109–124
Masopust D, Choo D, Vezys V et al (2010) Dynamic T cell migration program provides resident memory within intestinal epithelium. J Exp Med 207:553–564

Masopust D, Schenkel JM (2013) The integration of T cell migration, differentiation and function. Nat Rev Immunol
Masopust D, Vezys V, Marzo AL, Lefrancois L (2001) Preferential localization of effector memory cells in nonlymphoid tissue. Science 291:2413–2417
Miller JD, van der Most RG, Akondy RS et al (2008) Human effector and memory CD8+ T cell responses to smallpox and yellow fever vaccines. Immunity 28:710–722
Monteleone G, Pallone F, MacDonald TT (2011) Emerging immunological targets in inflammatory bowel disease. Curr Opin Pharmacol 11:640–645
Moon JJ, Chu HH, Pepper M, McSorley SJ, Jameson SC, Kedl RM, Jenkins MK (2007) Naive CD4(+) T cell frequency varies for different epitopes and predicts repertoire diversity and response magnitude. Immunity 27:203–213
Mueller SN, Gebhardt T, Carbone FR, Heath WR (2013) Memory T cell subsets, migration patterns, and tissue residence. Annu Rev Immunol 31:137–161
Newell EW, Klein LO, Yu W, Davis MM (2009) Simultaneous detection of many T-cell specificities using combinatorial tetramer staining. Nat Methods 6:497–499
Newell EW, Sigal N, Bendall SC, Nolan GP, Davis MM (2012) Cytometry by time-of-flight shows combinatorial cytokine expression and virus-specific cell niches within a continuum of CD8(+) T cell phenotypes. Immunity
Newell EW, Sigal N, Nair N, Kidd BA, Greenberg HB, Davis MM (2013) Combinatorial tetramer staining and mass cytometry analysis facilitate T-cell epitope mapping and characterization. Nat Biotechnol
Nikolich-Zugich J, Li G, Uhrlaub JL, Renkema KR, Smithey MJ (2012) Age-related changes in CD8 T cell homeostasis and immunity to infection. Semin Immunol 24:356–364
Nolz JC, Starbeck-Miller GR, Harty JT (2011) Naive, effector and memory CD8 T-cell trafficking: parallels and distinctions. Immunotherapy 3:1223–1233
Pawelec G, Koch S, Franceschi C, Wikby A (2006) Human immunosenescence: does it have an infectious component? Ann N Y Acad Sci 1067:56–65
Peretz Y, He Z, Shi Y et al (2012) CD160 and PD-1 co-expression on HIV-specific CD8 T cells defines a subset with advanced dysfunction. PLoS Pathog 8:e1002840
Peters PJ, Borst J, Oorschot V et al (1991) Cytotoxic T lymphocyte granules are secretory lysosomes, containing both perforin and granzymes. J Exp Med 173:1099–1109
Ravkov EV, Myrick CM, Altman JD (2003) Immediate early effector functions of virus-specific CD8+CCR7+ memory cells in humans defined by HLA and CC chemokine ligand 19 tetramers. J Immunol 170:2461–2468
Romero P, Zippelius A, Kurth I, et al (2007) Four functionally distinct populations of human effector-memory CD8+ T lymphocytes. J Immunol 178:4112–4119
Sakuishi K, Apetoh L, Sullivan JM, Blazar BR, Kuchroo VK, Anderson AC (2010) Targeting Tim-3 and PD-1 pathways to reverse T cell exhaustion and restore anti-tumor immunity. J Exp Med 207:2187–2194
Sallusto F, Geginat J, Lanzavecchia A (2004) Central memory and effector memory T cell subsets: function, generation, and maintenance. Annu Rev Immunol 22:745–763
Sallusto F, Lenig D, Forster R, Lipp M, Lanzavecchia A (1999) Two subsets of memory T lymphocytes with distinct homing potentials and effector functions. Nature 401:708–712
Sandalova E, Laccabue D, Boni C et al (2010) Contribution of herpesvirus specific CD8 T cells to anti-viral T cell response in humans. PLoS Pathog 6:e1001051
Sanders ME, Makgoba MW, Sharrow SO, Stephany D, Springer TA, Young HA, Shaw S (1988) Human memory T lymphocytes express increased levels of three cell adhesion molecules (LFA-3, CD2, and LFA-1) and three other molecules (UCHL1, CDw29, and Pgp-1) and have enhanced IFN-gamma production. J Immunol 140:1401–1407
Saule P, Trauet J, Dutriez V, Lekeux V, Dessaint JP, Labalette M (2006) Accumulation of memory T cells from childhood to old age: central and effector memory cells in CD4(+) versus effector memory and terminally differentiated memory cells in CD8(+) compartment. Mech Ageing Dev 127:274–281

Schluns KS, Kieper WC, Jameson SC, Lefrancois L (2000) Interleukin-7 mediates the homeostasis of naive and memory CD8 T cells in vivo. Nat Immunol 1:426–432
Schwartz RH (2003) T cell anergy. Annu Rev Immunol 21:305–334
Scott-Browne JP, White J, Kappler JW, Gapin L, Marrack P (2009) Germline-encoded amino acids in the alphabeta T-cell receptor control thymic selection. Nature 458:1043–1046
Seder RA, Darrah PA, Roederer M (2008) T-cell quality in memory and protection: implications for vaccine design. Nat Rev Immunol 8:247–258
Sharma P, Wagner K, Wolchok JD, Allison JP (2011) Novel cancer immunotherapy agents with survival benefit: recent successes and next steps. Nat Rev Cancer 11:805–812
Sigmundsdottir H, Butcher EC (2008) Environmental cues, dendritic cells and the programming of tissue-selective lymphocyte trafficking. Nat Immunol 9:981–987
Sprent J, Surh CD (2011) Normal T cell homeostasis: the conversion of naive cells into memory-phenotype cells. Nat Immunol 12:478–484
Starr TK, Jameson SC, Hogquist KA (2003) Positive and negative selection of T cells. Annu Rev Immunol 21:139–176
Su LF, Kidd BA, Han A, Kotzin JJ, Davis MM (2013) Virus-specific CD4(+) memory-phenotype T cells are abundant in unexposed adults. Immunity 38:373–383
Toebes M, Coccoris M, Bins A et al (2006) Design and use of conditional MHC class I ligands. Nat Med 12:246–251
Van Rhijn I, Kasmar A, de Jong A et al (2013) A conserved human T cell population targets mycobacterial antigens presented by CD1b. Nat Immunol 14:706–713
Wakim LM, Woodward-Davis A, Bevan MJ (2010) Memory T cells persisting within the brain after local infection show functional adaptations to their tissue of residence. Proc Natl Acad Sci USA 107:17872–17879
Waldrop SL, Pitcher CJ, Peterson DM, Maino VC, Picker LJ (1997) Determination of antigen-specific memory/effector CD4+ T cell frequencies by flow cytometry: evidence for a novel, antigen-specific homeostatic mechanism in HIV-associated immunodeficiency. J Clin Invest 99:1739–1750
Wherry EJ (2011) T cell exhaustion. Nat Immunol 12:492–499
Wherry EJ, Blattman JN, Murali-Krishna K, van der Most R, Ahmed R (2003) Viral persistence alters CD8 T-cell immunodominance and tissue distribution and results in distinct stages of functional impairment. J Virol 77:4911–4927
Wilkinson TM, Li CK, Chui CS et al (2012) Preexisting influenza-specific CD4+ T cells correlate with disease protection against influenza challenge in humans. Nat Med 18:274–280
Yuan J, Gnjatic S, Li H et al (2008) CTLA-4 blockade enhances polyfunctional NY-ESO-1 specific T cell responses in metastatic melanoma patients with clinical benefit. Proc Natl Acad Sci USA 105:20410–20415
Zhu J, Hladik F, Woodward A et al (2009) Persistence of HIV-1 receptor-positive cells after HSV-2 reactivation is a potential mechanism for increased HIV-1 acquisition. Nat Med 15:886–892

# Mass Cytometry to Decipher the Mechanism of Nongenetic Drug Resistance in Cancer

**Harris G. Fienberg and Garry P. Nolan**

**Abstract** Nongenetic resistance has recently been described as a major impediment to effective cancer therapy. Nongenetic resistance is challenging to study since it occurs nonuniformly, even in cell lines, and can involve the interplay of multiple survival pathways. Until recently, no technology allowed measurement of large-scale alterations in survival pathways with single-cell resolution. Mass cytometry, a flow-based technique in which the activation of up to 50 proteins can be measured simultaneously in single-cell, now provides the ability to examine nongenetic resistance on the functional level on a cell-by-cell basis. The application of mass cytometry, in combination with new bioinformatic techniques, will allow fundamental questions on nongenetic resistance to be addressed: Is resistance caused by selection of cells with a pre-existing survival phenotype or induction of a survival program? Which survival pathways are necessary for nongenetic resistance and how do they interact? Currently, mass cytometry is being used to investigate the mechanism of nongenetic resistance to TRAIL-induced apoptosis. The approaches being developed to understand resistance to TRAIL will likely be applied to elucidate the mechanisms of nongenetic resistance broadly and in the clinic.

## Contents



H. G. Fienberg · G. P. Nolan (✉)
Baxter Laboratory in Stem Cell Biology, Department of Microbiology and Immunology, Stanford University, Stanford, CA 94305, USA
e-mail: gnolan@stanford.edu

Current Topics in Microbiology and Immunology (2014) 377: 85–94
DOI: 10.1007/82_2014_365


Published Online: 28 February 2014

# 1 Introduction

The advent of targeted therapeutic agents was hailed as a major break through in the treatment of cancer. However, in most cases, initial promise is thwarted by the rapid development drug resistance. For targeted therapies to have sustained patient benefit, it will be necessary to understand and develop methods to combat resistance mechanisms. Canonically, resistance has been understood as a genetic process driven solely by mutations in therapeutic targets and associated regulators (Hanahan and Weinberg 2000). Nongenetic resistance has recently been implicated as an additional and significant hindrance to therapeutic efficacy (Marusyk et al. 2012). A host of mechanisms including upregulation of survival proteins, epigenetic modifications, regulated "noise" in gene expression, and selective activation of drug pumps have been proposed to account for nongenetic resistance (Brock et al. 2009; Sharma et al. 2010; Pisco et al. 2013). These mechanisms may operate in a coordinated manner or be activated selectively in different cancers or in response to distinct biological or chemical challenges. An ongoing question is whether nongenetic resistance is driven by the induction or by selection of survival mechanisms.

Theoretic and computational studies on the nature of the epigenetic (nongenetic) landscape suggest that either inductive or selective mechanisms could be at play in maintaining resistance (Waddington 1957; Pujadas and Feinberg 2012). In a conceptualization of the network state of the cell, known as a Waddington Landscape, the cellular signaling network state is depicted as a gravity well in which there is an interplay between unstable network states (visualized as being on the side of a hill) and stable network states (visualized as being on the floor of a valley). In the case of induction of a survival program, cells with unstable network states (hill states) are induced by the perturbation to assume more differentiated, stable network states (valley states) that encode a survival phenotype. In the selective case, cells in the unperturbed state are already distributed into survival network states (valley states) by virtue of nongenetic heterogeneity, and the perturbation does not significantly alter the diversity of survival phenotypes. Rather, the addition of a perturbation acts like a strainer to select cells with the network states that are resilient to the perturbation (Fig. 1).

# 2 Difficulties in the Study of Nongenetic Resistance

The study of the interplay between induction and selection of resistance states has been challenging. In order to examine the induction of survival pathways, it is necessary to measure a large number of proteins simultaneously, and in order to measure the selection of cells with survival phenotypes, it is necessary to use a single-cell approach. Before the advent of mass cytometry, it was not possible to

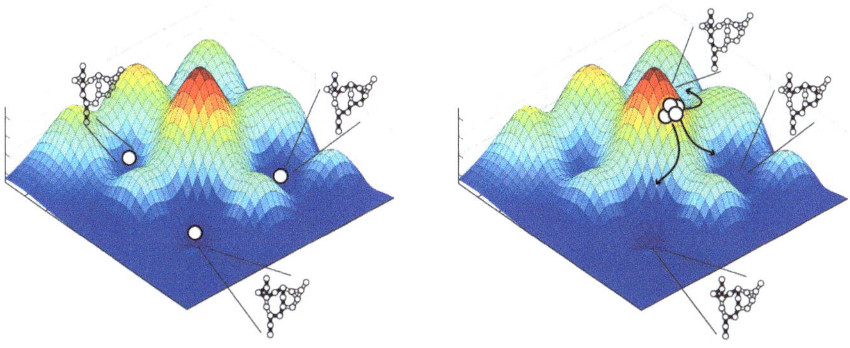


**Fig. 1** Depiction of Waddington Landscape schematics for selection and induction of survival mechanisms. Peaks correspond to unstable network states; valleys correspond to stable network states encoding survival phenotype. Chain link diagrams represent signaling networks transducing survival signal via different pathways in each example. (*left*) If nongenetic resistance is achieved by selection of cells with resistant network states, then cells in the basal state are already distributed into survival network states (valley states) and the perturbation acts as sieve to select the resistant phenotypes. (*right*) Alternatively, if resistance is achieved by induction of survival network states, cells with unstable network states (hill states) are induced by the perturbation to assume more differentiated, resistant network states (valley states)

simultaneously detect a large number of proteins with single-cell resolution. Early studies seeking to understand the contribution of induction to nongenetic resistance utilized bulk methods to measure the large number of proteins necessary to assess network state. These methods include reverse-phase protein microarray (Lee et al. 2012), receptor tyrosine kinase arrays (Rodrik-Outmezguine et al. 2011), quantitative western blotting (Lee et al. 2012), and multiplexed kinase inhibitor beads/mass spectroscopy (MIB/MS). In MIB/MS beads are used to enrich receptor tyrosine kinases from a population of cells and then the activity state of individual kinases is assayed by MALDI TOF/TOF (Oppermann et al. 2009; Duncan et al. 2012).

Studies on the contribution of selection to nongenetic resistance have employed the tools of single-cell analysis, chiefly fluorescent microscopy and fluorescently activated cell sorting (FACS), which are practically limited to measuring no more than a half dozen intracellular proteins simultaneously except for a few unique cases (Spencer and Sorger 2011; Sachs 2005) (Fig. 2).

In order to understand the relative contributions of induction and selection of survival phenotypes to nongenetic resistance, it is necessary to use a technology, such as mass cytometry, that is able to measure the activation states of a large number of proteins in response to drug exposure, simultaneously, and on a cell-by-cell basis. Mass cytometry is a flow-based technique in which the activation of up to 50 proteins can be measured simultaneously in a single cell. Data from 1,000 cells can be collected per second, making the analysis of millions of cells possible in a routine workflow (Bendall et al. 2011).

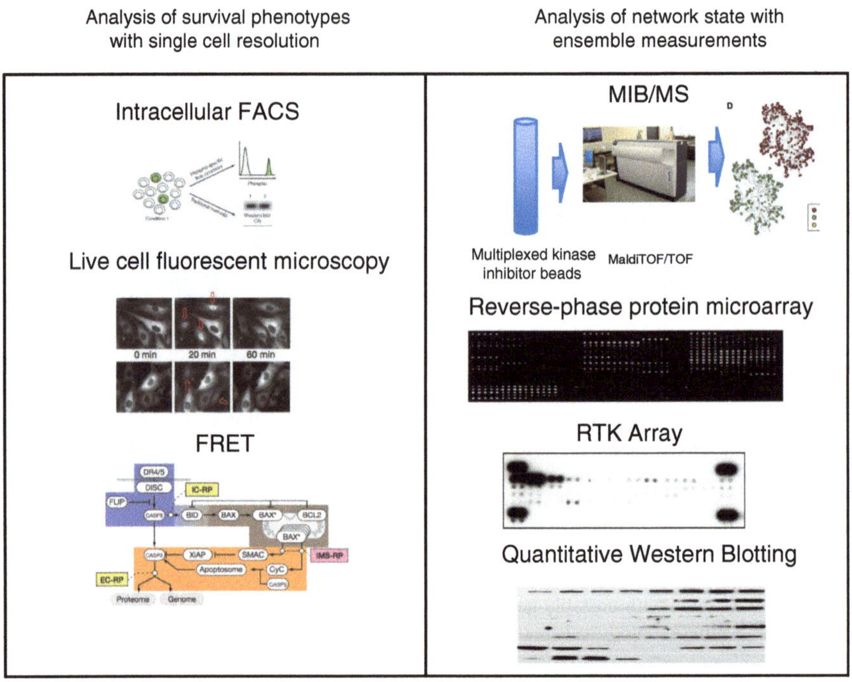


**Fig. 2** Pre-dating mass cytometry there was no method to examine a large number of proteins (20+) with single-cell resolution. A variety of single-cell methods have been used to examine selection of cells with survival phenotypes. These include intracellular FACS and phosphoflow (Krutzik and Nolan 2003; Spencer et al. 2009), live cell fluorescent microscopy (Cohen et al. 2008), and Förster resonance energy transfer (FRET) (Albeck et al. 2008). A number of distinct methods that take ensemble (i.e., bulk) measurements have been used to assay a large number of proteins to examine network state. These methods include MIB/MS (Oppermann et al. 2009; Duncan et al. 2012), reverse-phase protein microarrays (Lee et al. 2012), receptor tyrosine kinase arrays (Rodrik-Outmezguine et al. 2011), and quantitative western blotting

## 3 Features of Induction and Selection of Nongenetic Resistance

Induction of nongenetic resistance is most commonly referred to in the literature as network rewiring or compensatory signaling (Ryoo et al. 2004; Lee et al. 2012). In a basic example of induction, blockage of one survival pathway leads to the enhancement of an alternate survival pathway, which correlates with the presence of a resistant population (Rodrik-Outmezguine et al. 2011). Induction can act systematically and involve changes in the activation states of dozens of proteins throughout the cell leading to the increased prevalence of new network configurations that circumvents therapeutic intervention (Duncan et al. 2012).

The mechanism underlying induction of new phenotypic states has yet to be described and studies that have claimed to have induced a new network configuration may potentially be selecting for a subpopulation with a network state that is distinct from the modal network state before perturbation. Since these studies have not been performed with single-cell resolution, it is difficult to ascertain if a selective mechanism is involved. One study, however, has demonstrated that by sequential, but not simultaneous, application of targeted inhibitors it is possible to suppress the survivor phenotype, suggesting that it is possible to induce a change in the network state that affects phenotypic outcome (Lee et al. 2012).

Overall, the literature describing selection of nongenetic resistance demonstrates that particular network states that exist before perturbation can confer nongenetic resistance. Resistant cells can be detected based on their basal mitogenic protein expression before addition of therapy, implicating selection as a driver of resistance through nongenetic heterogeneity (Slack et al. 2008). It has been established that a diverse distribution of protein levels and protein activation potential can lead to increased survival (Slack et al. 2008; Singh et al. 2010). In normal tissues, variability in network state can be regulated and contribute to a population in which any single-cell may be susceptible to therapeutic intervention, but the population as a whole is unlikely to be completely ablated (Yuan et al. 2011). Diversity generating mechanisms may be directly selected for, as is seen in a number of other areas, such as ecology and microbiology, in which diversity has been shown to lead to system-wide robustness (Raser 2004; Flynn et al. 2011).

Diversity in general may help promote survival by allowing for the selection of cells occupying widespread survival niches. In addition, cells with a cancer stem cell phenotype can enable survival by occupying a refractory, quiescent state. The addition of chemotherapy induces apoptosis in faster growing cells and, therefore, selects for the quiescent cells (Reya et al. 2001). Surviving cells are positive for the cancer stem cell markers CD133 and CD24 and can be ablated by the use of histone deacetylase inhibitors, suggesting that DNA modifications enable the existence of this population (Sharma et al. 2010).

Induction of survival phenotypes can play a role in maintaining resistance after selection, suggesting that these two paradigms are not mutually exclusive in promoting nongenetic resistance. Several transcriptional regulators are selectively activated in surviving cells after treatment with the chemotherapeutic camptothecin, which activates downstream survival proteins (Cohen et al. 2008). Following death receptor activation with TRAIL and the initial selection of cells with survivor phenotypes, there is evidence that upstream survival signaling is induced in a subset of the survivors to maintain resistance (Spencer et al. 2009).

# 4 Use of Mass Cytometry to Decipher the Role of Inductive and Selective Mechanisms in Nongenetic Drug Resistance

Mass cytometry overcomes the limitations of previous technologies by providing parametric breadth with single-cell resolution. It is uniquely suited to analyzing complex, network processes that occur heterogeneously. Mass cytometry is currently being used by the Nolan group to determine whether selection or induction of survival network states underlies nongenetic drug tolerance to TRAIL-induced apoptosis. TRAIL-induced apoptosis is an ideal model process for understanding the mechanisms underlying nongenetic resistance. It occurs as the result of a directed and physiologically relevant stimulus, therefore, its induction does not involve significant off-target and complicating effects, as chemotherapy and even targeted therapy can. Most of the effectors of TRAIL-induced apoptosis are well studied, and sensitive markers of TRAIL induction are available (Johnstone et al. 2008). Finally, cell cycle state and positional effects have been ruled out as drivers of nongenetic resistance in this system (Spencer et al. 2009; Flusberg et al. 2013; Flusberg and Sorger 2013). A panel of antibodies against over 3 dozen proteins implicated in survival response from TRAIL-induced apoptosis, including multiple markers of activation in the MAPK, JNK, p38, and NF$\kappa$B pathways have been validated. These pathways in combination with TRAIL are necessary for survival (Sah et al. 2003; Ohtsuka et al. 2003; Frese et al. 2003; Weldon et al. 2004).

Mass cytometry is a destructive technology, so survivors cannot be tracked to determine whether a particular network state is induced by a perturbation. However, the network state of cells can be traced on the subpopulation level using bioinformatics clustering approaches. New algorithms, such as viSNE, allow high-dimensional data to be compressed into a 2-D map of phenotypic space (Amir et al. 2013). Therefore, one can determine whether the signaling state of survivors fits within the survival niche of cells in the basal state, suggesting a selection paradigm, or whether the survivors create a novel niche that did not exist in the basal state, suggesting an induction paradigm (Fig. 3).

The network state of survivors can also be probed to determine whether induction or selection of the survivor phenotype underlies survival. The simultaneous measurement of the functional states of dozens of proteins in single-cell allows construction of network maps using the correlations of each protein in the network. Using straightforward statistical methods, how network features vary with survival can be determined. If the perturbation induces selective pressure that result in a survival phenotype, those perturbations that increase apoptosis (and decrease survival) should show a high level of net change in the network. In this case, the perturbation effectively acts as a sieve, and a small subset of cells with network states that differ substantially from the modal cell network state survive. Conversely, if the survival phenotype is induced, then conditions with the greatest

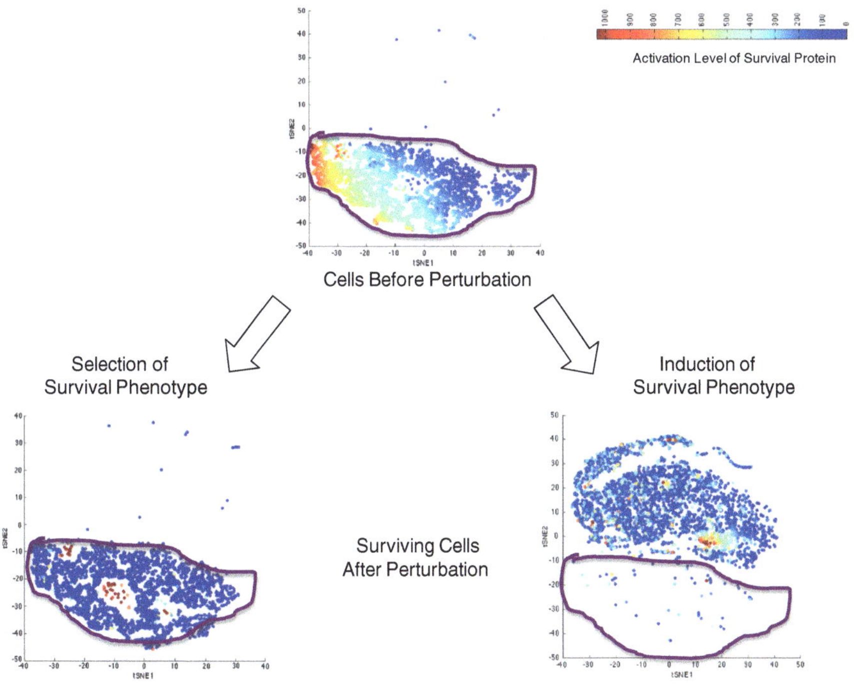


**Fig. 3** Simulated data showing network state of survivors in viSNE plot. Cells are clustered based on functional alterations in 20 survival proteins. Each circle represents a single-cell. Cells in close proximity possess similar network states. Cells are colored based on activation level of one of the survival proteins. Area within purple line represents signaling state of cells before perturbation. If nongenetic resistance results from a selection mechanism, the surviving cells will fit within the signaling space of cells in the basal state. If nongenetic resistance results from an inductive mechanism, the surviving cells will occupy a signaling space that is distinct from the signaling space of cells in the basal state. If nongenetic resistance results from a hybrid mechanism (not pictured), cells will occupy both the pre-existing signaling space and a new signaling space

level of net change in the network should correlate with perturbations that do not increase apoptosis. In this situation, the perturbation acts as a road block, forcing cells to drive further out of their way (i.e., alter their network state more profoundly) in order to reach the same destination (a survival network state).

By combining bioinformatics clustering to map the phenotypic state of the survivors and network deconvolution methods to dissect the level of network alteration in the survivors, it will be possible to determine whether induction, selection, or some combination of the two mechanisms, supports nongenetic resistance to TRAIL. Furthermore, these methods represent a generalizable approach that can be expanded to other systems to determine how these mechanisms of nongenetic resistance interact more broadly.

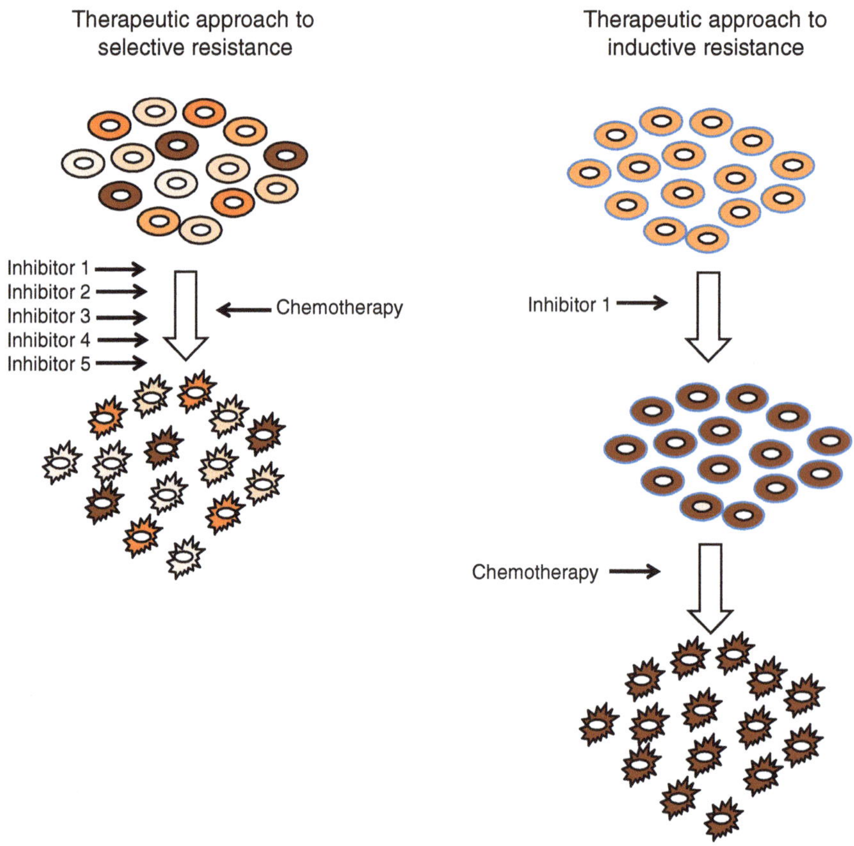


**Fig. 4** Schematics representing methods to exploit selective or inductive mechanisms of nongenetic resistance with targeted therapy. In cases in which selective resistance occurs, low-dose multidrug combinations could be used to decrease the number of network states that are resistant to therapy. In cases of inductive resistance, pretreatment with targeted inhibitors could induce cells to assume a network configuration that is more susceptible to follow-on therapy

## 5 Conclusion

Despite great promise, no TRAIL-based therapy is used clinically, even though at least nine therapeutics targeting TRAIL having entered clinical trials (Newsom-Davis et al. 2009; Dimberg et al. 2012). Nongenetic drug tolerance may be a significant contributor to the lackluster clinical trial results. In cancers such as chronic myelogenous leukemia, nongenetic drug tolerance appears to act in concert with genetic mutations to render therapy ineffective (Okabe et al. 2008; Brock et al. 2009). Disentangling whether induction or selection underlies nongenetic resistance in a particular cancer may allow design of smarter and more effective cancer therapies (Fig. 4). If a survival phenotype is induced, then it may be possible to push the survivors into a less refractory state by the application of targeted inhibitors prior to the use of chemotherapy (Lee et al. 2012). If a selective process is at work, the simultaneous application of poly-specific inhibitors could be used to narrow the possible range of survival niches and lead to an ablation of resistant populations.

## References

Albeck JG, Burke JM, Aldridge BB et al (2008) Quantitative analysis of pathways controlling extrinsic apoptosis in single cells. Mol Cell 30:11–25. doi:10.1016/j.molcel.2008.02.012

Amir E-AD, Davis KL, Tadmor MD et al (2013) viSNE enables visualization of high dimensional single-cell data and reveals phenotypic heterogeneity of leukemia. Nat Biotechnol 31:545–552. doi:10.1038/nbt.2594

Bendall SC, Simonds EF, Qiu P et al (2011) Single-cell mass cytometry of differential immune and drug responses across a human hematopoietic continuum. Science 332:687–696. doi:10.1126/science.1198704

Brock A, Chang H, Huang S (2009) Non-genetic heterogeneity—a mutation-independent driving force for the somatic evolution of tumours. Nature 10:1–7

Cohen AA, Geva-Zatorsky N, Eden E et al (2008) Dynamic proteomics of individual cancer cells in response to a drug. Science 322:1–6

Dimberg LY, Anderson CK, Camidge R et al (2012) On the TRAIL to successful cancer therapy? Predicting and counteracting resistance against TRAIL-based therapeutics. Oncogene 32:1341–1350. doi:10.1038/onc.2012.164

Duncan JS, Whittle MC, Nakamura K et al (2012) Dynamic reprogramming of the kinome in response to targeted MEK inhibition in triple-negative breast cancer. Cell 149:307–321. doi:10.1016/j.cell.2012.02.053

Flusberg DA, Roux J, Spencer SL, Sorger PK (2013) Cells surviving fractional killing by TRAIL exhibit transient but sustainable resistance and inflammatory phenotypes. Mol Biol Cell. doi:10.1091/mbc.E12-10-0737

Flusberg DA, Sorger PK (2013) Modulating cell-to-cell variability and sensitivity to death ligands by co-drugging. Phys Biol 10:035002. doi:10.1088/1478-3975/10/3/035002

Flynn DFB, Mirotchnick N, Jain M et al (2011) Functional and phylogenetic diversity as predictors of biodiversity–ecosystem-function relationships. Ecology 92:1573–1581

Frese S, Pirnia F, Miescher D et al (2003) PG490-mediated sensitization of lung cancer cells to Apo2L/TRAIL-induced apoptosis requires activation of ERK2. Oncogene 22:5427–5435. doi:10.1038/sj.onc.1206842

Hanahan D, Weinberg RA (2000) The hallmarks of cancer. Cell 100:57–70

Johnstone RW, Frew AJ, Smyth MJ (2008) The TRAIL apoptotic pathway incancer onset, progression and therapy. Nat Rev Cancer 8:782–798. doi: 10.1038/nrc2465

Krutzik PO, Nolan GP (2003) Intracellular phospho-protein staining techniques for flow cytometry: monitoring single cell signaling events. Cytometry 55A:61–70. doi:10.1002/cyto.a.10072

Lee MJ, Ye AS, Gardino AK et al (2012) Sequential application of anticancer drugs enhances cell death by rewiring apoptotic signaling networks. Cell 149:780–794. doi:10.1016/j.cell.2012.03.031

Marusyk A, Almendro V, Polyak K (2012) Intra-tumour heterogeneity: a looking glass for cancer? Nat Publ Group 12:323–334. doi:10.1038/nrc3261

Newsom-Davis T, Prieske S, Walczak H (2009) Is TRAIL the holy grail of cancer therapy? Apoptosis 14:607–623. doi:10.1007/s10495-009-0321-2

Ohtsuka T, Buchsbaum D, Oliver P et al (2003) Synergistic induction of tumor cell apoptosis by death receptor antibody and chemotherapy agent through JNK/p38 and mitochondrial death pathway. Oncogene 22:2034–2044. doi:10.1038/sj.onc.1206290

Okabe S, Tauchi T, Ohyashiki K (2008) Characteristics of dasatinib- and imatinib-resistant chronic myelogenous leukemia cells. Clin Cancer Res 14:6181–6186. doi:10.1158/1078-0432.CCR-08-0461

Oppermann FS, Gnad F, Olsen JV et al (2009) Large-scale proteomics analysis of the human kinome. Mol Cell Proteomics 8:1751–1764. doi:10.1074/mcp.M800588-MCP200

Pisco AO, Brock A, Zhou J et al (2013) Non-Darwinian dynamics in therapy-induced cancer drug resistance. Nat Commun 4:2467. doi:10.1038/ncomms3467

Pujadas E, Feinberg AP (2012) Regulated noise in the epigenetic landscape of development and disease. Cell 148:1123–1131. doi:10.1016/j.cell.2012.02.045

Raser JM (2004) Control of stochasticity in eukaryotic gene expression. Science 304:1811–1814. doi:10.1126/science.1098641

Reya T, Morrison SJ, Clarke MF, Weissman IL (2001) Stem cells, cancer, and cancer stem cells : abstract : nature. Nature 414:105–111. doi:10.1038/35102167

Rodrik-Outmezguine VS, Chandarlapaty S, Pagano NC et al (2011) mTOR kinase inhibition causes feedback-dependent biphasic regulation of AKT signaling. Cancer Discov 1(3)

Ryoo HD, Gorenc T, Steller H (2004) Apoptotic cells can induce compensatory cell proliferation through the JNK and the wingless signaling pathways. Dev Cell 7:491–501. doi:10.1016/j.devcel.2004.08.019

Sachs K (2005) Causal protein-signaling networks derived from multiparameter single-cell data. Science 308:523–529. doi:10.1126/science.1105809

Sah NK, Munshi A, Kurland JF et al (2003) Translation inhibitors sensitize prostate cancer cells to apoptosis induced by tumor necrosis factor-related apoptosis-inducing ligand (TRAIL) by activating c-Jun N-terminal kinase. J Biol Chem 278:20593–20602. doi:10.1074/jbc.M211010200

Sharma SV, Lee DY, Li B et al (2010) A chromatin-mediated reversible drug-tolerant state in cancer cell subpopulations. Cell 141:69–80. doi:10.1016/j.cell.2010.02.027

Singh DK, Ku C-J, Wichaidit C et al (2010) Patterns of basal signaling heterogeneity can distinguish cellular populations with different drug sensitivities. Mol Syst Biol 6:1–10. doi:10.1038/msb.2010.22

Slack MD, Martinez ED, Wu LF, Altschuler SJ (2008) Characterizing heterogeneous cellular responses to perturbations. Proc Natl Acad Sci 105(49):1–6

Spencer SL, Sorger PK (2011) Measuring and modeling apoptosis in single cells. Cell 144:926–939. doi:10.1016/j.cell.2011.03.002

Spencer SL, Gaudet S, Albeck JG et al (2009) Non-genetic origins of cell-to-cell variability in TRAIL-induced apoptosis. Nature 459:428–432. doi:10.1038/nature08012

Waddington CH (1957) The strategy of the genes. A discussion of some aspects of theoretical biology. With an appendix by Kacser H. Allen & Unwinpp, London, pp. ix. 262

Weldon CB, Parker AP, Patten D et al (2004) Sensitization of apoptotically-resistant breast carcinoma cells to TNF and TRAIL by inhibition of p38 mitogen-activated protein kinase signaling. Int J Oncol 24:1473–1480

Yuan TL, Wulf G, Burga L, Cantley LC (2011) Cell-to-cell variability in PI3 K protein level regulates PI3 K-AKT pathway activity in cell populations. Curr Biol 21:173–183. doi:10.1016/j.cub.2010.12.047

# A Practical Guide to Multiplexed Mass Cytometry

**Nevena Zivanovic, Andrea Jacobs and Bernd Bodenmiller**

**Abstract** Recent advances in inductively coupled plasma mass spectrometry (ICP-MS) as applied in mass cytometry, enabled its broad applicability to life science research. Mass cytometry enables the high-dimensional characterization of cellular systems by simultaneously measuring dozens of metal isotope reporter labeled antibodies bound to cell components. With the ability to simultaneously interrogate an unprecedented number of molecular components on a per cell basis, it offers the possibility to gain better understanding of single cell biology in heterogeneous samples. To upscale this single cell information to screening approaches by mass cytometry, a cell-based multiplexing technique, called mass-tag cellular barcoding (MCB), was developed. MCB enables the simultaneous analysis of multiple cell samples by using $n$ metal ion tags to multiplex up to $2^n$ samples. Different mass tag combinations are used to label individual cell samples with a unique mass barcode that allows multiple samples to be combined and immunostained together for a single analysis on the mass cytometer. Taken together, MCB enables increased sample throughput, reduces antibody consumption, and increases the overall data quality. In this chapter, we describe the MCB to array the samples in a 96-well format that allows for medium-scale profiling/screening experiments to be run on a standard mass cytometer.

## Contents



N. Zivanovic · A. Jacobs · B. Bodenmiller (✉)
Institute of Molecular Life Sciences, University of Zürich, 8057 Zürich, Switzerland
e-mail: bernd.bodenmiller@imls.uzh.ch

Current Topics in Microbiology and Immunology (2014) 377: 95–109
DOI: 10.1007/82_2013_335

Published Online: 6 August 2013

## 1 Introduction

Mass cytometry is a recently developed technology platform that allows for high-content, multiparametric analysis of single cells in complex biological systems (Ornatsky et al. 2010). Using mass cytometry, one can simultaneously interrogate signaling components, cell cycle state, cell viability, DNA content, calcium flux, and many other cellular markers (Bendall et al. 2011). Coupled to simultaneous cell surface marker profiling, these unique single cell states can be put into the context of complex heterogeneous cell populations such as those encountered in human tissue samples. There are a wide range of possible applications for mass cytometry including drug screening, patient profiling, biomarker discovery, and time-course analysis to name just a few (Bodenmiller et al. 2012). Until recently, however, the number of samples that could be analyzed by mass cytometry was limited due to the protocols used for sample preparation and the mass cytometric measurement itself: Immune-staining over multiple samples was heterogeneous, the antibody reagent costs were high, and the fluidics systems on the mass cytometer allowed only a low sample throughput.

To address these issues, the multiplexing approach employed in fluorescent cell barcoding (FCB) (Krutzik and Nolan 2006; Krutzik et al. 2008) was adapted for use in mass cytometry in a method of cell multiplexing that we call mass-tag cellular barcoding (MCB) (Bodenmiller et al. 2012). MCB enables the simultaneous analysis of multiple samples in a single measurement (Bodenmiller et al. 2012). To accomplish this, individual cell samples are labeled with a unique combination of mass tags before being combined into a single sample. The pooled sample is stained with a single antibody mix and analyzed in one run on the mass cytometer. Measured cells are thereafter assigned to the corresponding source sample based on their unique "mass barcode" signatures (Bodenmiller et al. 2012). By mixing samples prior to staining, antibody consumption is typically reduced 30- to 50-fold. In addition, the overall data quality is increased through the

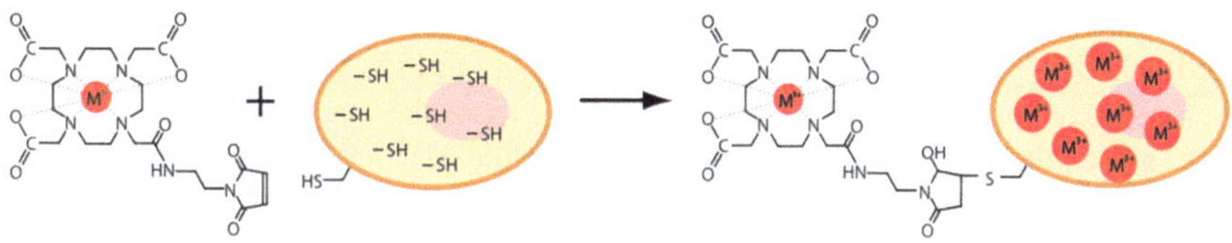

**Fig. 1** Mass-tag cellular barcoding. Cells are covalently labeled with the bifunctional compound mDOTA. This compound can be loaded with lanthanide (III) isotope ions and reacts covalently with cellular thiol groups through the maleimide moiety. Reproduced from Bodenmiller et al. (2012)

identical processing of control and all samples of interest, which eliminates pipetting errors and staining variation and reduces artifacts due to instrument variations. Finally, the sample analysis throughput is greatly increased as the cleaning step of the fluidics system can be omitted. Taken together, MCB enables increased sample throughput, reduces antibody consumption, and increases the overall data quality (Bodenmiller et al. 2012), thereby enabling medium-scale profiling/screening experiments to be run on a standard mass cytometer. In this book chapter, we describe MCB procedures that can be used for any cell samples of interest.

## 2 Reagents

In MCB, cells are labeled with unique signatures or "barcodes". This is accomplished by using mass tags—bifunctional chelating reagents that are able to strongly chelate specific metal ions and that can be covalently bound to a cell. For the MCB procedure described in this chapter, the molecule 1,4,7,10-tetraazacyclododecane-1,4,7-tris–acetic acid 10-maleimide ethylacetamide (maleimido-mono-amido-DOTA or mDOTA) is used as the bifunctional compound. The DOTA moiety readily chelates rare earth metal lanthanide (III) ions with a $K_d$ of $\sim 10^{-16}$, and the maleimide moiety rapidly reacts covalently with cellular thiol groups as illustrated in Fig. 1 (Bodenmiller et al. 2012). Since the reactive bifunctional reagent is covalently attached to the cell, unreacted molecules can be readily removed by washing, the mass-tagged samples are stable over multiple sample processing steps, and the samples can be stored a week or more without cross contamination of tags between cells within the multiplexed sample.

Using binary combinations of $n$ metal ion tags it is possible to multiplex up to $2^n$ samples. By using seven or nine metal tags one can multiplex 96 or 384 samples, respectively (Bodenmiller et al. 2012). For mass cytometry, the lanthanide series of transition metal elements are used as they are not normally present in biological samples, and many different stable isotopes are available at high purity. In addition, their uniform +3 oxidation state gives great flexibility in choosing metal isotopes without the need of using different chelating chemistries. Although all rare-earth metals are amenable to MCB, highly pure isotopes at the high end of the lanthanide series (e.g., Yb171, Yb172, Yb174, Yb175, Yb176) are preferred as their use avoids false positive signals resulting from the isotope oxides (M + 16) that can result from oxidation of an isotope in the plasma (Ornatsky et al. 2010; Tanner et al. 2013). Since lanthanide metal isotopes are used in a combinatorial fashion to barcode large number of samples, individual mass barcodes will differ in the number of metals within a unique combination. In order to maintain the same ratio of maleimide groups to thiol groups across the barcoding plate, the amount of mDOTA per well is equalized across the plate by addition of mDOTA-conjugated metal isotopes that are not detected by mass cytometer (i.e., that do not interfere with the measurement channels) but that have similar analytical properties (e.g., Ga (III)). Metals used for this purpose must have a low mass so that their signals are excluded from the mass range used for detection and a +3 oxidation state such that they are efficiently chelated by mDOTA.

# 3 Preparing mDOTA-Lanthanide (III) DMSO Stocks for CyTOF® Barcoding

## *3.1 Materials*

- mDOTA (Macrocyclics, #B272)
- Lanthanide (III) metal isotopes as chloride salts (DVS sciences)
- Gallium (III) as chloride salt (Sigma-Aldrich, #427128)
- 20 mM acetate buffer, pH 5.5
- Dimethyl sulfoxide (DMSO, Sigma-Aldrich, # D2438).

## *3.2 Method*

The metal (III)-mDOTA complex should be prepared at 2:1 molar ratio of mDOTA:metal, in order to reduce or eliminate unchelated metal in the barcoding reagent.

1. Weigh out 2 mg of solid mDOTA into a 1.5 mL microcentrifuge tube. Spin to the bottom of the tube for about 5 s.
2. Dissolve the metal (III) salt stock solution in 20 mM acetate buffer, pH 5.5 to a concentration of 25 mM.
3. Dissolve the solid mDOTA at 41.5 mg/mL (to 50 mM final concentration) in 20 mM acetate buffer pH 5.5 containing 25 mM metal (III) salt.
4. Solid may not dissolve immediately. Vortex to dissolve it completely.
5. Snap freeze by placing in liquid nitrogen for 10–20 s.

6. Pierce the tube cap with a small hole, and transfer tube quickly to a lyophilization bell jar.
7. Lyophilize overnight or as necessary depending on the total volume.

***Note:*** *Make sure sample does not thaw before putting on vacuum—this will lead to boiling.*

***Note:*** *When removing the sample from the lyophilizer, the dried pellet may fly out of the microcentrifuge tube due to static electricity and/or the air rushing in when venting the lyophilization jar. One solution is to poke a hole in the tube cap with a wide-gauge needle.*

8. Spin tube to collect all powder at the bottom of the microcentrifuge tube.
9. Dissolve the white pellet in DMSO, aliquot, and freeze for long-term storage at −20 °C. We typically use 10 mM stocks. Store at −20 °C.

# 4 Titrating Barcoding Reagents

Barcoding is well suited for analysis of both cultured immortalized cells and primary cell samples. The reaction between the maleimide group of the barcoding reagent and the cellular thiol groups is essentially complete in 15 min at room temperature. The optimal concentration of barcoding reagent depends on the total number of thiol groups in the sample, which is a function of cell number and cell volume. We recommend that similar cell numbers (or better, a similar number of thiol groups) be used in each well for the barcoding experiments. The concentration of each mass-tag barcoding reagent must be titrated for the experimentally desired number of target cells. MCB is typically used in experiments where cells have been crosslinked (fixed) and permeabilized for analysis of intracellular antigens such as protein phosphorylation sites (Bodenmiller et al. 2012). This workflow is preferable because the thiol-reactive reagents used find many more targets inside the cell as compared to the cell surface.

## *4.1 Materials*

- Live cells
- PBS (Sigma-Aldrich, #P4417-100TAB), at 4 °C
- Doubly distilled water (ddH$_2$O)
- Cell Staining Media (CSM): PBS, pH 7.4. with 0.5 % bovine serum albumin and 0.02 % sodium azide, at 4 °C

- 1.6 % paraformaldehyde (PFA): dilute 16 % PFA (Electron Microscopy Sciences, #15710) 1:10 in PBS
- Ice-cold methanol, HPLC grade (Sigma-Aldrich, #646377-1L)
- FACS tubes (BD Falcon, #352008)
- Filter-cap FACS tubes (BD Falcon, #352235)
- Inject-F 1-mL syringes (Fisher, #S7510-1)
- Iridium metallointercalator working solution: PBS, pH 7.4, with 1.6 % PFA and 0.02 % iridium metallointercalator (DVS Sciences, #Inter-1X-natIr).

## *4.2 Method*

### 4.2.1 Preparation of Formaldehyde Crosslinked and Permeabilized Cells

1. Bring cells into suspension and add 1.6 % PFA; incubate for 10 min at room temperature.
2. Centrifuge cells at 600 × *g* for 5 min.
3. Remove supernatant completely.
4. Vortex vigorously to resuspend all cells, add ice-cold methanol to a concentration of ~1 × $10^6$ cells/mL, incubate for 10 min at 4 °C to permeabilize cells.

***Note:*** *Optional pause point. Cells can be stored long-term in methanol at −80 °C.*

5. Take six FACS tubes per dilution series for each barcoding reagent to be titrated.
6. Add 3 mL CSM and 1 mL cell suspension (1 × $10^6$ cells) to each tube.
7. Centrifuge at 600 × *g* for 5 min at 4 °C.
8. Remove supernatant.
9. Resuspend cell pellet in 3 mL PBS and centrifuge at 600 × *g* for 5 min at 4 °C.
10. Remove supernatant and resuspend cell pellet in 1 mL PBS.

### 4.2.2 Titration of Barcoding Reagents

Prepare a six-step dilution series of each barcoding reagent; dilutions are made in DMSO.

1. Dilute the 10 mM stock barcoding reagent to 500 μM with DMSO (i.e., 5 μL 10 mM stock barcoding reagent plus 95 μL DMSO).
2. Use the 500 μM barcoding reagent to prepare a titration series according the following scheme:

| Concentration of barcoding reagent [μM] | 250 | 50 | 10 | 2 | 0.4 | 0.08 |
|---|---|---|---|---|---|---|
| Volume 500 μM barcoding reagent [μL] | 50 | | | | | |
| Volume DMSO [μL] | 50 | 80 | 80 | 80 | 80 | 80 |

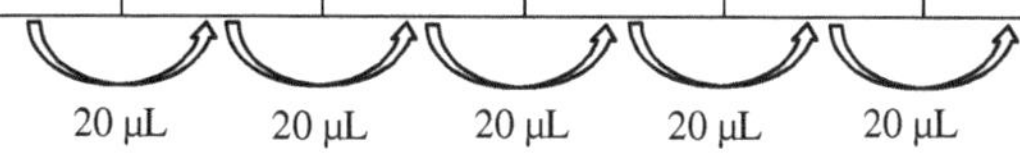


3. Add 4 μL barcoding reagent from each tube of the titration series to 1 mL cell suspension in PBS and vortex vigorously.

| Final concentration of barcoding reagent [nM] | 1000 | 200 | 40 | 8 | 1.6 | 0.32 |
|---|---|---|---|---|---|---|

***Note:*** *DMSO tends to crystallize when pipetted into ice-cold (aqueous) solutions; therefore, it is critical to vortex immediately after pipetting the barcoding reagent into the cell suspension.*

4. Incubate for 30 min at room temperature. Keep cells in suspension by vortexing intermittently.
5. Add 3 mL CSM and centrifuge at 600 × *g* for 5 min at 4 °C.
6. Remove supernatant.
7. Repeat washing step twice.
8. Remove supernatant.
9. Remove supernatant and add appropriate volume of Ir-metallointercalator working solution (1.6 % PFA in PBS with 1:5000 metallointercalator) to each tube.
10. Use 1 mL of 1:5000 metallointercalator for ~1–2 × $10^6$ cells.
11. Incubate for 20 min at room temperature.

***Note:*** *Optional pause point. Cells can be stored in metallointercalator working solution overnight at 4 °C.*

12. Add 3 mL CSM and centrifuge at 600 × *g* for 5 min at 4 °C.
13. Remove supernatant.
14. Repeat washing step once.
15. Add 3 mL PBS and centrifuge at 600 × *g* for 5 min at 4 °C.
16. Remove supernatant.

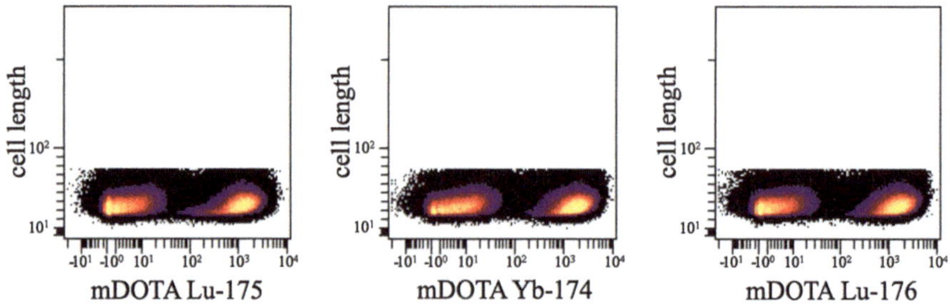


**Fig. 2** Example of optimally titrated barcoding reagents

17. Add 3 mL $ddH_2O$ and centrifuge at 600 × *g* for 5 min at 4 °C.
18. Remove supernatant.
19. Resuspend cells at ~1 × $10^6$ cells/mL with $ddH_2O$.
20. Filter through filter-cap FACS tube, and transfer to 1 mL syringe.
21. Acquire sample data on CyTOF®.

***Note:*** *Always begin the measurements with the LOWEST CONCENTRATION and proceed from lowest concentration to highest. If the signal becomes too high, the instrument can be damaged. Do not continue if the signal in a sample is >32 k ion counts per cell event.*

#### 4.2.3 Data Analysis

Analyze acquired data, and determine the optimal concentration for making the barcoding plates, taking into account signal intensity and separation between cells positive and negative for barcoding reagent as shown in Fig. 2.

## 5 Barcoding Plate Layout

Barcoding is performed by adding barcoding reagents in a combinatorial, binary fashion to each individual cell sample of interest. For convenience, a barcoding plate is made in a 96-well plate containing different combinations of barcoding reagents in each well. The pattern of binary combinations of the barcoding reagents distributed over the plate is called the barcoding key. The number of combinations depends on the number of different barcoding reagents (i.e., different metal isotopes used for making barcoding reagents). Additionally, in order to address the metal content variability across the plate, and avoid potential signal shifts, the metal content between wells and therefore samples must be equalized. Thus, in parallel to addition of a given barcoding reagent to each well (i.e., Lu175), the metal content is balanced by the addition of same amount of

**Table 1** Concentrations of reagents used for preparation of barcoding plate

| | DOTA:Lanthanide complexes | Reagent working solution | Master plate | Barcoding plate | Final concentration |
|---|---|---|---|---|---|
| Barcoding reagent | mDOTA:Tm169 | 400 μM | 40 μM | 4 μM | 40 nM |
| | mDOTA:Er170 | 400 μM | 40 μM | 4 μM | 40 nM |
| | mDOTA:Yb171 | 400 μM | 40 μM | 4 μM | 40 nM |
| | mDOTA:Yb172 | 400 μM | 40 μM | 4 μM | 40 nM |
| | mDOTA:Yb174 | 400 μM | 40 μM | 4 μM | 40 nM |
| | mDOTA:Lu175 | 400 μM | 40 μM | 4 μM | 40 nM |
| | mDOTA:Yb176 | 400 μM | 40 μM | 4 μM | 40 nM |
| Equalization reagent | mDOTA:Ga70 | 400 μM | 40 μM | 4 μM | 40 nM |
| Volume | | 720 μL | 100 μL | 100 μL | 1000 μL |

equalization reagent to wells not containing the given barcoding reagent. Importantly, the principle of the protocol remains the same regardless of a barcoding key used for making the barcoding plates. Multiple barcoding plates can be prepared at the same time by preparing one "master" barcoding plate containing barcoding reagents at high concentration, which can be diluted into individual barcoding plates. In the following section, we describe the generation of a 96-well barcoding plates similar to the setup described in Bodenmiller et al. (2012).

## 5.1 Materials

- Thermo-fast 96 Detection Plate (Thermo, #AB1400-L)
- Aluminum sealing film (Axygen, #PCR-AS-200)
- DMSO (Sigma-Aldrich, # D2438)
- Barcoding reagents
- Equalization reagent

## 5.2 Method

### 5.2.1 Preparation of the Master Plate

The master plate will contain the mixed barcoding reagents at a 10-fold higher concentration than the concentrations needed for individual barcoding plates (Table 1). To generate the master plate, 10 μL of barcoding reagent working solution is pipetted into appropriate wells, and the reagent for equalization is pipetted in the wells not containing that barcoding reagent (Fig. 3). Equalization reagent is used in the same volume (10 μL) and in the same concentration as the barcoding reagent. This means that when seven barcoding reagents are used, the total volume in each well will be 100 μL: 70 μL barcoding reagent plus

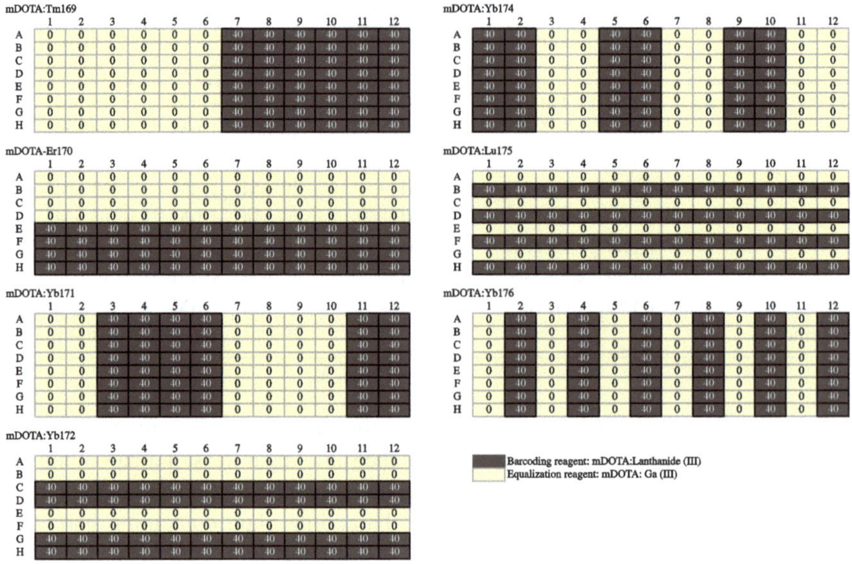


**Fig. 3** Layout of barcoding matrix used to encode 96 samples. Seven unique lanthanide isotopes were used to generate 128 combinations, enough to barcode each sample in 96-well plate. The concentrations of each of the seven lanthanide isotopes and that of the equalizing reagent are shown as they are distributed on the 96-well plate

equalization reagent and 30 μL DMSO. Working solutions of the barcoding reagents and equalizing reagents are prepared by diluting the stock solutions (10 mM) to the concentrations needed for the master plate preparation.

***Note:*** *In order to minimize pipetting errors, the barcoding key layout was designed to be pipetted using a multi-channel pipette. We advise that individual reagents are first pipetted in "patterns" suitable for the multi-channel pipette and at higher volume than it is required in "intermediate" 96-well plates, so that they can be easily transferred to the master plate using the multi-channel pipette.*

### 5.2.2 Preparation of Barcoding Plates from Master Plate

***Note:*** *Use a multi-channel pipette for preparing the barcoding plates to minimize pipetting errors.*

1. Pipette 90 μL DMSO into each well of nine 96-well plates.
2. Transfer 10 μL of barcoding reagent from the master plate to the appropriate wells in nine barcoding plates.
3. Mix gently by pipetting up and down.
4. Dilute the remaining 10 μL of barcoding reagent in the master plate by adding 90 μL of DMSO.
5. Seal barcoding plates with aluminum foil and store at −20 °C until used.

# 6 Validation of the MCB Master Plate

To validate the accuracy and robustness of the MCB method, it is essential to verify that the MCB method does not alter mass cytometry measurements or introduce artifacts. In addition, it is important to ensure that the pattern of the barcoding reagent distribution in the barcoding plate can be retrieved upon deconvolution of the data acquired on the mass cytometer. Typically, two types of cell populations differing in a component to be immunostained (e.g., stimulated vs. unstimulated cells) are stained and used to validate the MCB master plate. Two cell types are arranged in geometrical shapes across the 96-well plate to create a checkerboard or striped pattern. After 96-well multiplexing, mass cytometry analysis of a particular antigen abundance is performed. Deconvolution should yield the expected pattern of cell sample distribution on the 96-well plate (Fig. 4) (Bodenmiller et al. 2012). In addition, the data acquired for multiplexed cell samples should be compared to non-multiplexed samples to ensure similar behavior (Bodenmiller et al. 2012; Krutzik and Nolan 2006; Krutzik et al. 2008).

## *6.1 Materials*

- $20 \times 10^6$ crosslinked and permeabilized cells from stimulated and unstimulated samples
- Metal-tagged antibodies (e.g., DVS Sciences or self-conjugated)
- CSM at 4 °C
- Iridium metallointercalator working solution (DVS Sciences, #Inter-1X-natIr)
- $ddH_2O$
- PBS (Sigma Aldrich, #P4417-100TAB)
- Cluster tubes (Corning, #4418)
- FACS tubes (BD Falcon, #352008)
- Filter cap FACS tubes (BD Falcon, #352235)
- Inject-F 1 mL syringes (Fisher, #S7510-1).

## *6.2 Method*

1. Aliquot cells into the cluster tubes by pipetting $0.2 \times 10^6$ cells/tube (200 μL of $10^6$ cell/mL suspension).
2. Organize cell samples according to one of the patterns shown in Fig. 4.
3. Add CSM to a volume of 1 mL; centrifuge at 600 × *g* for 5 min at 4 °C.
4. Remove supernatant.
5. Resuspend cells in 1 mL PBS and centrifuge at 600 × *g* for 5 min at 4 °C.
6. Remove supernatant.
7. Resuspend cells in 500 μL PBS.

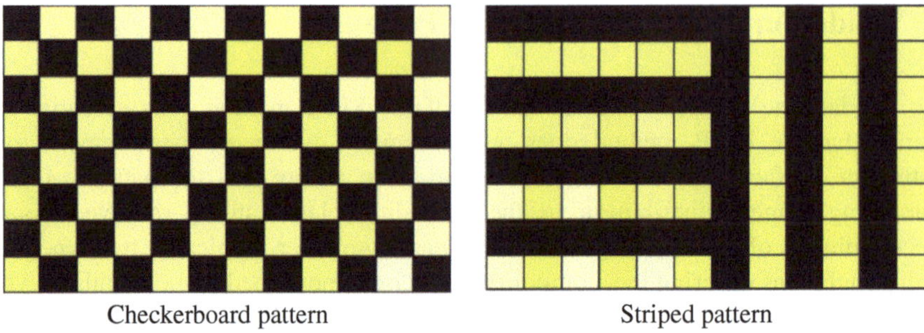


**Fig. 4** Checkerboard and striped patterns of cell sample distribution used for validation of MCB analysis. Reproduced from Bodenmiller et al. (2012)

8. Add 5 μL of barcoding reagent from the barcoding plate and mix immediately by pipetting.
9. Incubate for 30 min at room temperature.
10. Add 500 μL CSM and centrifuge at 600 × *g* for 5 min at 4 °C.
11. Remove supernatant.
12. Repeat washing step twice.
13. Mix all cells from the checkerboard or striped pattern into a single FACS tube.
14. Perform immunostaining by resuspending the cell pellets in 300 μL of an appropriate antibody mix.
15. Mix gently and incubate of 60 min at room temperature.
16. Add CSM to a volume of 3 mL.
17. Centrifuge at 600 × *g* for 5 min at 4 °C.
18. Remove supernatant.
19. Add 3 mL CSM and centrifuge at 600 × *g* for 5 min at 4 °C.
20. Remove supernatant.
21. Proceed with Ir-metallointercalation (1:5000) protocol and cell preparation for mass cytometric analysis (see above).
22. Acquire sample data on CyTOF® (unlimited cell number, acquisition time of 1,200 s).
23. Deconvolute data.

## 7 Data Deconvolution

An essential step for multiplexed mass cytometry is the ability to retrieve the information on each original individual sample based on their barcoded mass signatures. This is achieved by using Boolean gating of barcoded cells in biaxial plots of mDOTA-lanthanide (III) channels, to define cell populations containing defined barcoding signatures as presented in the Fig. 5 (Bodenmiller et al. 2012). Boolean

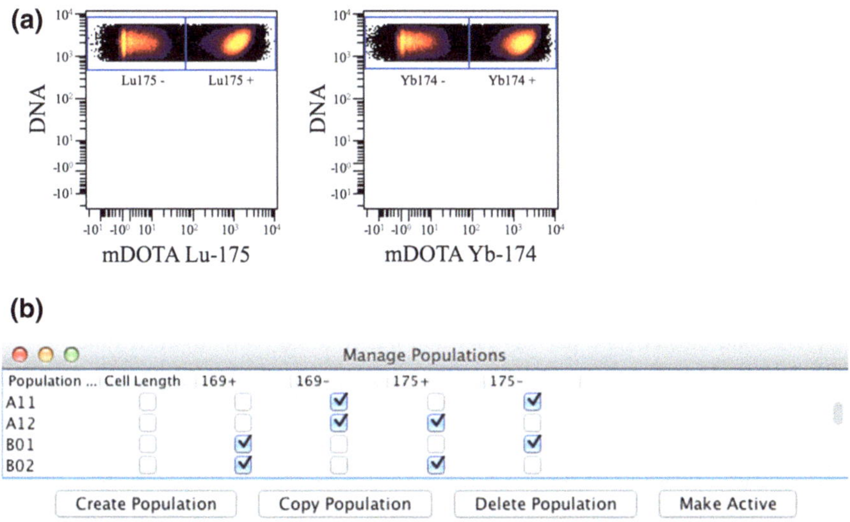


**Fig. 5** Boolean gating. **a** Density dot biaxial plots of barcoded cells are shown with DNA content on the y axis and barcoding channel on the x axis. Cells positive and negative for a given channel are indicated. **b** Standard gating techniques are combined with Boolean algebraic operations (AND, NOT, OR) to generate Boolean gates and, subsequently, to assign to each cell sample a particular barcode signature and position on a barcoding plate

algebraic operations (AND, NOT, and OR) are combined with standard cell population gating techniques to generate Boolean gates in flow cytometry (Brown 2003). By creating multiple combinations of operations, any gating structure can be implemented during analysis to define cell populations containing a particular barcoding signature and subsequently to assign cells to their original sample.

## 8 Barcoding and Intracellular Staining of Cell Samples

Here, we describe a straightforward procedure for combining MCB with immunostaining of cell samples. The protocol can easily adapt to lab-specific requirements. Important for a successful barcoding is the cell permeabilization and the removal of unbound barcoding reagent.

### *8.1 Materials*

- Crosslinked and permeabilized cells (see above)
- PBS (Sigma Aldrich, #P4417-100TAB)
- dd$H_2O$

- Cluster tubes (Corning, #4418)
- FACS tubes (BD Falcon, #352008)
- Filter cap FACS tubes (BD Falcon, #352235)
- Inject-F 1 mL syringes (Fisher, #S7510-1)
- CSM at 4 °C
- Iridium metallointercalator working solution (DVS Sciences, #Inter-1X-natIr).

## *8.2 Method*

1. Use one cluster tube per sample; add 0.2–1 × $10^6$ cells to each tube.
2. Add CSM to 1 mL.
3. Centrifuge at 600 × *g* for 5 min at 4 °C.
4. Remove supernatant.
5. Resuspend cell pellet in 1 mL PBS and centrifuge at 600 × *g* for 5 min at 4 °C.
6. Remove supernatant.
7. Resuspend cell pellet in 500 μL PBS.
8. Add 5 μL of barcoding reagent and mix immediately by pipetting.
9. Incubate for 30 min at room temperature. Keep cells in suspension by vortexing intermittently.
10. Add 500 μL CSM and centrifuge at 600 × *g* for 5 min at 4 °C.
11. Remove supernatant.
12. Add 1 mL CSM and centrifuge at 600 × *g* for 5 min at 4 °C.
13. Remove supernatant.
14. Repeat washing step twice.
15. Combine all cell samples into a single FACS tube.
16. Centrifuge at 600 × *g* for 5 min at 4 °C.
17. Perform immunostaining by resuspending the cell pellets in 300 μL of an appropriate antibody mix.
18. Mix gently and incubate for 60 min at room temperature.
19. Add CSM to 3 mL and centrifuge at 600 × *g* for 5 min at 4 °C.
20. Remove supernatant.
21. Add 3 mL CSM and centrifuge at 600 × *g* for 5 min at 4 °C.
22. Remove supernatant.
23. Proceed with Ir-metallointercalation (1:5000) protocol and cell preparation for mass cytometric analysis (see above).
24. Acquire sample data on CyTOF® (unlimited cell number, acquisition time of 1,200 s).
25. Deconvolute data.

## References

Bendall SC, Simonds EF, Qiu P, Amir el-AD, Krutzik PO, Finck R, Bruggner RV, Melamed R, Trejo A, Ornatsky OI, Balderas RS, Plevritis SK, Sachs K, Pe'er D, Tanner SD, Nolan GP (2011) Single-cell mass cytometry of differential immune and drug responses across a human hematopoietic continuum. Science 332:687–696

Bodenmiller B, Zunder ER, Finck R, Chen TJ, Savig ES, Bruggner RV, Simonds EF, Bendall SC, Sachs K, Krutzik PO, Nolan GP (2012) Multiplexed mass cytometry profiling of cellular states perturbed by small-molecule regulators. Nat Biotechnol 30:858–867

Brown FM (2003) Boolean reasoning: the logic of Boolean equations. Dover Publications, New York

Krutzik PO, Nolan GP (2006) Fluorescent cell barcoding in flow cytometry allows high-throughput drug screening and signaling profiling. Nat Methods 3:361–368

Krutzik PO, Crane JM, Clutter MR, Nolan GP (2008) High-content single-cell drug screening with phosphospecific flow cytometry. Nat Chem Biol 4:132–142

Ornatsky O, Bandura D, Baranov V, Nitz M, Winnik MA, Tanner S (2010) Highly multiparametric analysis by mass cytometry. J Immunol Methods 361:1–20

Tanner SD, Baranov VI, Ornatsky OI, Bandura DR, George TC (2013) An introduction to mass cytometry: fundamentals and applications. Cancer Immunol Immunother 62:955–965

# Analysis of Protein Interactions in situ by Proximity Ligation Assays

**Björn Koos, Linda Andersson, Carl-Magnus Clausson, Karin Grannas, Axel Klaesson, Gaëlle Cane and Ola Söderberg**

**Abstract** The fate of the cell is governed by interactions among proteins, nucleic acids, and other biomolecules. It is vital to look at these interactions in a cellular environment if we want to increase our understanding of cellular processes. Herein we will describe how the in situ proximity ligation assay (in situ PLA) can be used to visualize protein interactions in fixed cells and tissues. In situ PLA is a novel technique that uses DNA, together with DNA modifying processes such as ligation, cleavage, and polymerization, as tools to create surrogate markers for protein interactions of interest. Different in situ PLA designs make it possible not only to detect protein–protein interactions but also post-translational modifications and interactions of proteins with nucleic acids. Flexibility in DNA probe design and the multitude of different DNA modifying enzymes provide the basis for modifications of the method to make it suitable to use in many applications. Furthermore, examples of how in situ PLA can be combined with other methods for a comprehensive view of the cellular activity status are discussed.

## Contents



B. Koos · L. Andersson · C.-M. Clausson · K. Grannas · A. Klaesson · G. Cane ·
O. Söderberg (✉)
Department of Immunology, Genetics and Pathology Science for Life Laboratory, BMC, Uppsala University, SE-751 85 Uppsala, Sweden
e-mail: ola.soderberg@igp.uu.se

Current Topics in Microbiology and Immunology (2014) 377: 111–126 
DOI: 10.1007/82_2013_334

Published Online: 7 August 2013

# 1 Introduction

Proteins are involved in most, if not all, processes in the cell through interactions with other protein-nucleic acid interaction, and other types of biomolecules. Understanding the vital processes of the cell requires methods to monitor protein interactions. In order to achieve this, probes that target the interacting proteins and a way to detect this binding are needed. The most commonly used probe to target proteins is the antibody, which has been used for protein detection for over half a century (Coons et al. 1942). Fluorophores or enzymes attached to the antibodies are used as reporters to visualize the presence of bound antibodies and, therefore, the protein of interest. The analysis of protein interactions is more complicated as signal generation is dependent on dual recognition by combining capture and detection; examples are ELISA, co-immunoprecipitation, and Förster (or bioluminescence) resonance energy transfer (FRET/BRET). An alternative approach is to let the proximal binding of two probes facilitate the formation of a reporter molecule. By using DNA as a reporter molecule, the enzymatic reactions of DNA modification (e.g., cleavage, ligation, and amplification) can be used to modify the reporters in a proximity-dependent manner. The proximity ligation assay (PLA) utilizes antibodies to which short single-stranded DNA oligonucleotides have been attached (Fredriksson et al. 2002; Gullberg et al. 2004) (Fig. 1). Upon binding of a pair of such PLA probes, the oligonucleotides will reside close together. After hybridization to a DNA oligonucleotide with regions complementary to both probes, the DNA strands on the antibodies can be ligated to form a bridge between the two antibodies. This newly joined DNA strand becomes a surrogate marker of the proximal binding of the PLA probes and can be amplified by PCR. The dual recognition by PLA probes required in the formation of a reporter molecule is an advantage, because it decreases non-specific signal as only the ligated reporter molecule can be amplified (Weibrecht et al. 2010).

Analysis of cell-to-cell communication and how individual cells are affected by their microenvironment is preferably performed in situ at a single-cell level. In order to achieve this, a DNA-amplification method other than PCR has to be used, so that the amplification product remains bound to the protein and does not diffuse away. To enable in situ analysis of protein interactions, we designed a PLA oligonucleotide system so that the proximal binding of a pair of PLA probes would template the ligation of two subsequently added DNA oligonucleotides into a circular DNA molecule (Soderberg et al. 2006) (Fig. 2). This DNA circle, the surrogate reporter for proximal binding, is then amplified using the highly processive phi29 DNA polymerase in a rolling circle amplification (RCA), primed by one of the PLA probes. The resulting RCA product will thus be an extension of the PLA probes, and hence physically linked to the interacting proteins via the antibodies. An RCA product consists of several hundred repetitive elements that can be labeled by hybridization of many fluorophore-conjugated short DNA oligonucleotides (detection oligonucleotides). This generates a very bright sub-μm-sized object that can be visualized by microscopy. The bright discrete RCA products

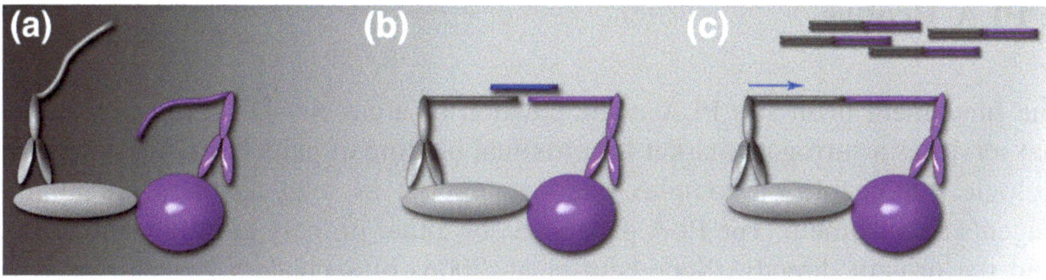


**Fig. 1** Schematic figure of proximity ligation assay using PCR for amplification. **a** Two PLA probes (antibodies with short, single-stranded DNA oligonucleotides attached) bind interacting proteins bringing their oligonucleotides into close proximity. **b** These oligonucleotides hybridize to a complementary bridging oligonucleotide (*blue*). **c** A ligase seals the nick between the two PLA probes, and the resulting DNA strand is amplified by PCR.

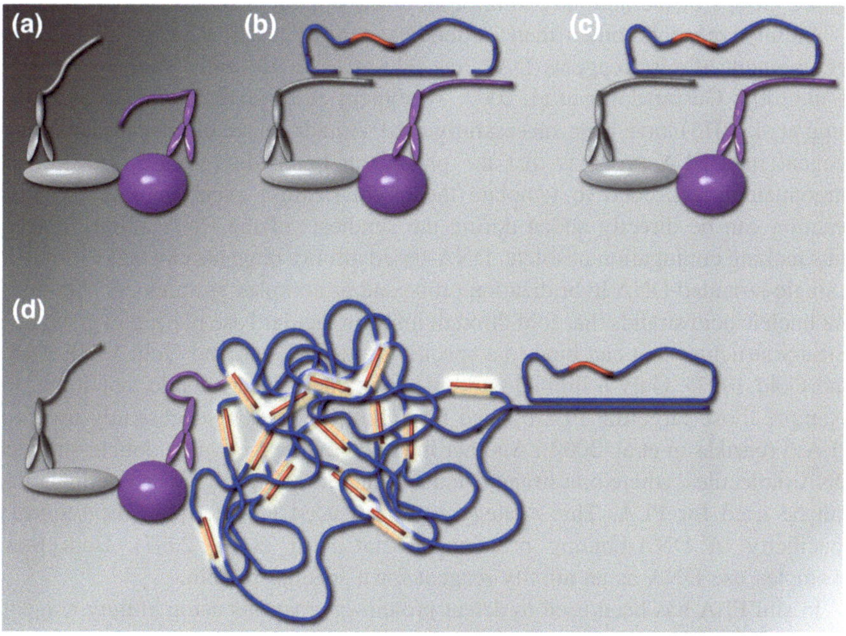


**Fig. 2** In situ proximity ligation assay (PLA) scheme using rolling circle amplification (RCA) to generate a localized amplification product. **a** The PLA probes bind to the two interacting proteins. **b** If the PLA probes bind in close proximity, two DNA oligonucleotides (circularization oligonucleotides) can hybridize to the probes. **c** The circularization oligonucleotides are ligated to form a circular DNA molecule. **d** The circular DNA molecule is amplified by phi 29 DNA polymerase resulting in a RCA product that can be visualized using fluorescently labeled oligonucleotides.

allow visualization and enumeration of single molecules by standard microscopy and facilitate enumerations by image analysis software (e.g., CellProfiler, http://www.cellprofiler.org) (Carpenter et al. 2006).

## 2 PLA Probes

The fundament of in situ PLA is the creation of a circular DNA oligonucleotide that serves as a surrogate marker for proximal binding of pairs of PLA probes. Any biological molecule or complex can be targeted as long as there are affinity reagents against them. The PLA probes can be either primary affinity reagents that bind the protein directly (Soderberg et al. 2006) or secondary reagents (Jarvius et al. 2007) that target pairs of antibodies bound to individual or interacting proteins. A PLA probe consists of an affinity reagent to which a DNA oligonucleotide has been bound. Commonly antibodies are used as affinity reagents. In order to convert an antibody into a PLA probe a chemical linker is required to attach the DNA oligonucleotides. Linkers such as succinimidyl-4-[N-maleimidomethyl]cyclohexane-1-carboxylate (SMCC) enable conjugation of primary amine groups of the antibody to thiol-modified oligonucleotides.

Affinity reagents other than antibodies can be used as PLA probes; the recombinant affinity reagents DARPins (Gu et al. 2013) and DNA (Fredriksson et al. 2002; Gustafsdottir et al. 2007; Weibrecht et al. 2012; Gomez et al. 2013; Jung et al. 2013) have been successfully used. An advantage of DNA as an affinity reagent in a PLA probe is that the probe can be readily synthesized and the oligonucleotide needed to template the circularization event later in the PLA reaction can be directly added during the synthesis of the DNA affinity reagent thus making conjugation obsolete. DNA-based affinity reagents can be as simple as a single-stranded DNA hybridization probe and as complex as aptamers. Aptamers are nucleic acid strands that fold through intra-molecular base pairing to generate a tertiary structure that can bind to a specific protein (Tuerk and Gold 1990; Burke and Gold 1997). Only a limited number of aptamers are available, but some, for example those targeting PDGF$\beta$ and thrombin, have been successfully used for PLA (Fredriksson et al. 2002). Another method development uses double-stranded DNA molecules, where one strand is longer than the other to give an overhang that can be used for PLA. This strategy has been used to determine the sequence specificity of DNA-binding proteins (Gustafsdottir et al. 2007). Both these examples use DNA as an affinity reagent for a specific protein.

In situ PLA has been used to detect protein interactions using affinity reagents against interacting proteins (Soderberg et al. 2006) and for detection of post-translational modifications (PTM) such as phosphorylations (Jarvius et al. 2007; Koos et al. 2009), glycosylations (Conze et al. 2010; Pinto et al. 2012) and SUMOylations (Matic et al. 2010; Sehat et al. 2010). Ectopically expressed protein-nucleic acid interaction can be monitored by using fusion tags to which the PLA probes are targeted (Gajadhar and Guha 2010; Gu et al. 2013). Moreover, DNA probes can be used to target nucleic acids to assay proximity between proteins and specific DNA or RNA elements. The easiest approach is to use a single-stranded DNA probe with an extension that serves as template for ligation (Fig. 3). Hybridization to genomic DNA by such a PLA probe requires that the genomic DNA is first rendered single stranded, by heat or enzymatic digestion (Weibrecht

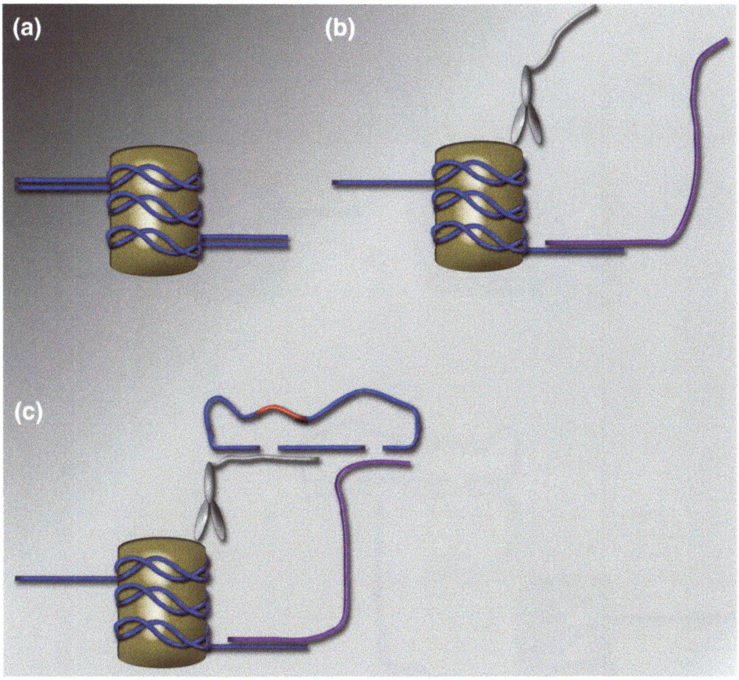


**Fig. 3** Protein-nucleic acid interaction detected by in situ PLA utilizing an antibody and a DNA-based probe. **a** Genomic DNA interacting with histones is double stranded and therefore needs to be digested or denatured by heat to become single stranded. **b** A DNA-based PLA probe (*purple*) hybridizes to the single-stranded genomic DNA, while an antibody-based PLA probe binds the histone. **c** The two probes act as templates for ligation of the circularization oligonucleotides making rolling circle amplification possible.

et al. 2012; Gomez et al. 2013). The process of generating single-stranded fragments of the genomic DNA may be exploited as tool to create potential PLA probes. The genomic fragments will contain protein bound to DNA, which may be targeted with an antibody-based PLA probe, while the single-stranded genomic DNA acts as a second PLA probe (Fig. 4). The circularization oligonucleotides must be custom-made for each genomic DNA sequence targeted.

To increase selectivity, the circularization oligonucleotides can be converted into a padlock-like probe. Padlock probes are oligonucleotides whose 5′ and 3′ ends are complementary to adjacent regions on a target DNA sequence (Nilsson et al. 1994). Hybridization of a padlock probe to its target brings the 3′ and 5′ ends together. The gap can then be sealed by a ligase, which will close the padlock probe into a circular conformation that can act as a template for RCA. The padlock-like probe can be opened after hybridization and ligation onto the genomic DNA. This opened probe will in turn target the PLA probe to create a new circle (Fig. 5). This approach also provides dual tag identification for increased selectivity.

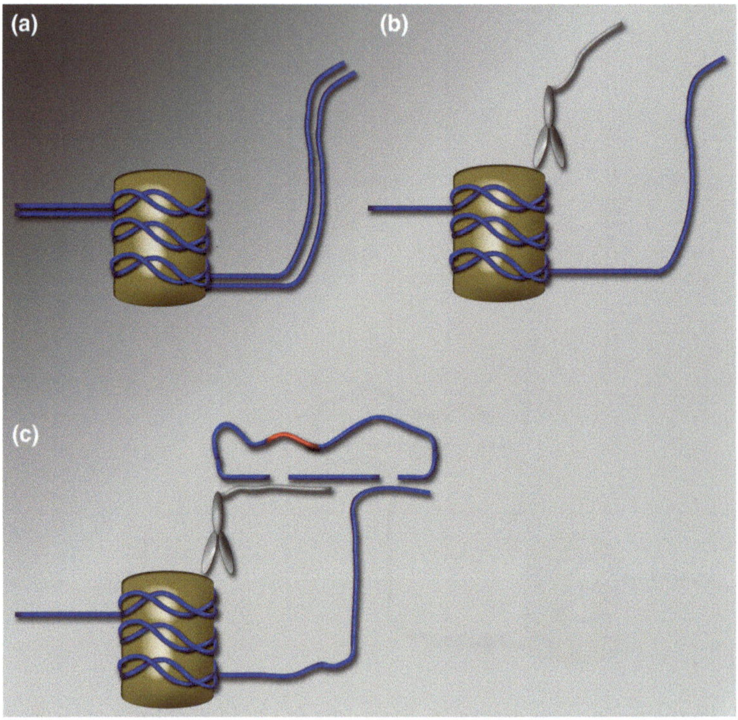


**Fig. 4** Protein-nucleic acid interaction detected by in situ PLA utilizing genomic DNA as PLA probe. **a** Genomic DNA is rendered single stranded. **b** An antibody-based PLA probe binds to the histone proximate to the single-stranded genomic DNA. **c** The antibody-based PLA probe and the genomic DNA act as a template for hybridization and ligation of the circularization oligonucleotides.

Detecting RNA–protein interactions is straightforward since RNA molecules are partially single stranded and therefore more readily available for hybridizing probes than is genomic DNA. However, targeting RNA has different challenges. Ligation of DNA nicks using RNA templates is rather inefficient since most ligases do not accept DNA/RNA heteroduplexes. Hybridization of multiply labeled tetravalent RNA imaging probes (MTRIPs, i.e., four biotinylated hybridizing oligonucleotides bound to Flag-tagged neutravidin) has been used to target mRNA in living cells. After fixation and permeabilization the Flag-tag and the RNA binding protein of interest can be targeted with PLA probes, which facilitate the detection of the protein-nucleic acid interaction (Jung et al. 2013).

## 3 Signal Generation

The proximal binding of a pair of PLA probes brings the oligonucleotide regions of the probes together to allow ligation to create a template for RCA. The distance requirement for the hybridization and ligation will depend on the size of the

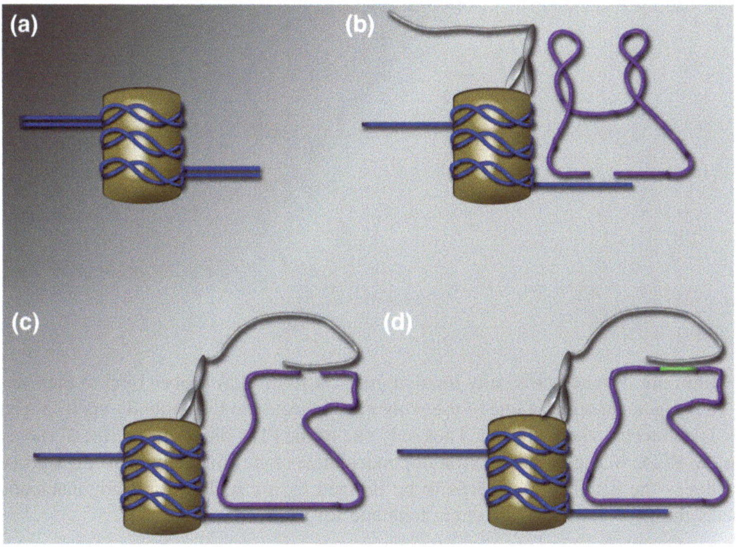


**Fig. 5** Protein-DNA interactions detected by in situ PLA utilizing a padlock-like probe. **a** Genomic DNA is rendered single stranded. **b** A padlock-like probe hybridizes to the single-stranded genomic, while an antibody-based PLA probe binds the histone. **c** Upon ligation of the padlock-like probe, it is enzymatically converted into a ligation substrate for the antibody-based PLA probe. **d** An oligonucleotide containing a tag sequence hybridizes to the antibody-based PLA probe. The padlock-like probe and the tag are ligated into a circle, which can be amplified by RCA.

affinity reagents and the lengths of the oligonucleotides. The size of an antibody is around 10 nm, and a nucleotide spans around 0.3 nm. The distance threshold that will allow ligation can be modified by using alternative affinity reagents or by changing the nucleotide sequence or orientation. The conjugation of the oligonucleotides to the affinity reagents is done at the 5′ end of the oligonucleotides, to provide a free 3′ end to prime the RCA. However, by conjugating one of the PLA probes to the 3′ end, the distance between the affinity reagents is decreased to a minimum (Fig. 6). With this design, only one of the PLA probes can initiate RCA, whereas both are required for the formation of a DNA circle (unpublished observations). By variation of length (i.e., number of DNA bases) this design could potentially be used to determine distance thresholds between epitopes.

Although the distance requirement between epitopes that allows the generation of a PLA signal is only a few nm, the size of the RCA product will be several hundred nm in diameter. One hour of amplification generates a product of around 100 kilobases (a 1,000-fold amplification of a 100 base circle), and many fluorophore-labeled detection oligonucleotides can hybridize to the product. An advantage of a high number of detection oligonucleotides is the very bright signal that is easily distinguishable from background fluorescence; this facilitates subsequent image analysis. The disadvantage is that the large size of the RCA

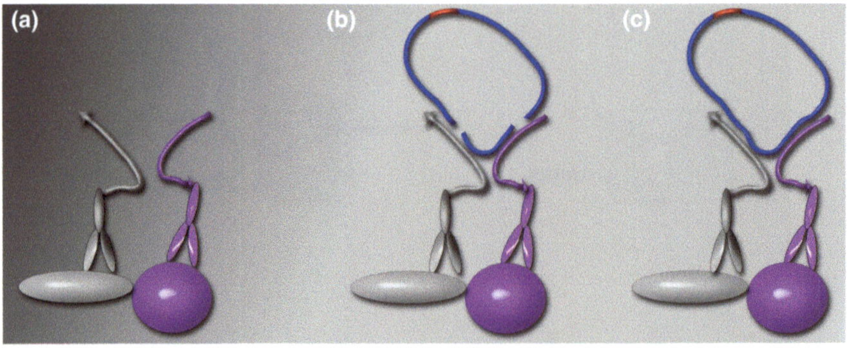


**Fig. 6** Minimum distance proximity ligation assay. **a** Both PLA probes bind to their respective targets. The oligonucleotide of one of the probes is connected to the antibody via its 3′ end, while the other is connected via the 5′ end. Thus only one of the PLA probes has a free 3′ end available to prime the RCA. **b** The circularization oligonucleotides hybridize to the two PLA probes. This design reduces the distance that needs to be bridged by a circularization oligonucleotide. **c** A ligase seals the gaps and creates a circle template for RCA.

products limits the dynamic range of the assay. If the RCA products are so abundant that they coalesce it is not possible to identify discrete signals, which is a requirement for digital enumeration. Quantification may also be achieved by measuring the intensity of the total signal; however, this intensity will not increase linearly, since at higher concentrations of RCA products mass-transport will put further constraints on the assay (i.e., the time to saturate all possible hybridization sites will increase) (Mocanu et al. 2011). Therefore enumeration of RCA products is preferable for quantification. To overcome the problem of coalescing signals, a combination of different circularization oligonucleotides, each equipped with a unique reporter tag, can be used to increase the dynamic range while allowing enumeration of individual signals. If tags are used at defined ratios, the generated circles, and thus subsequent RCA products, will contain the tags in the same ratio. When RCA products are visualized by fluorophore-labeled tag-specific detection oligonucleotides (i.e., a different fluorophore for each tag), differently colored RCA products will reflect the ratio between circularization oligonucleotides. Using this strategy, when the cells are saturated in one fluorescence channel, it is possible to switch to a different channel and enumerate the more sparse signals. Since the dilution factor of the circulation oligonucleotides is known, digital image analysis is possible (Clausson et al. 2011) (Fig. 7).

The use of different tag-sequences can also be employed for multiplexed read-out. By incorporating the tag sequence into the oligonucleotides of the PLA probes one can create a set of probes to monitor combinatorial events of many different proteins. We designed an oligonucleotide system with a set of circularization oligonucleotides used to create circles of all different combinations of tag sequences by incorporating the tag sequence in the middle of a PLA probe (Leuchowius et al. 2013) (Fig. 8). The RCA products carry information regarding

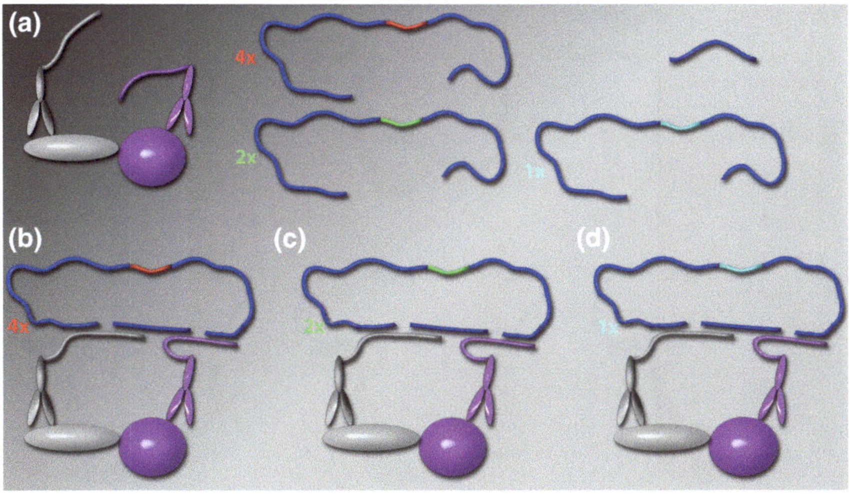


**Fig. 7** Increasing the dynamic range of in situ PLA read out. **a** After binding of the PLA probes to their targets, samples are incubated with a set of circularization oligonucleotides that differ only in the detection site for the fluorophore-labeled detection oligonucleotide. These probes are present at different concentrations (high amount, *red*; medium amount, *green*; and low amount, *cyan*) and can together with the shorter circularization oligonucleotide create a circle. **b** The *red* circularization oligonucleotide will occupy most of the PLA probes (since it has the highest concentration) and hence gives the most RCA products. This is useful if low abundance interactions are to be investigated. **c** The *green* oligonucleotide will give less RCA products than the *red* and can be used when the *red* signal is saturated. **d** The third tag (*cyan*) with the lowest concentration is used in cells with a very high amount of interactions, which would saturate the two other channels.

which PLA probe was involved in the creation of the DNA circle, and tags are identified by hybridization of tag-specific fluorophore-labeled detection oligonucleotides. For higher levels of multiplexing, sequential read-out of either a single fluorophore (Schubert et al. 2006) or a set of fluorophores (Goransson et al. 2009) can be used. For identification of complexes where both interaction partners can be replaced with another protein (in contrast to keeping one protein fixed as in Fig. 8), both PLA probes need to contain a tag that will be propagated into dual-labeled RCA products (Leuchowius et al. 2013).

Creation of circular DNA ligation products as reporter molecules of proximity is not limited to detection of pairwise interactions. To visualize proximity of three proteins, three different PLA probes must be designed so that the formation of a circular ligation product will need to involve all of them (Soderberg et al. 2006; Zieba et al. 2012) (Fig. 9). An assay has been developed for solution-phase PLA that requires recognition of five antibodies in order to create a PLA product (Tavoosidana et al. 2011). The limit of how many ligation events are possible in the formation of a PLA product is probably much higher. The efficiency of detecting proximity between multiple proteins will decrease as the number of PLA probes increases due to the fact that not all epitopes will be available and occupied

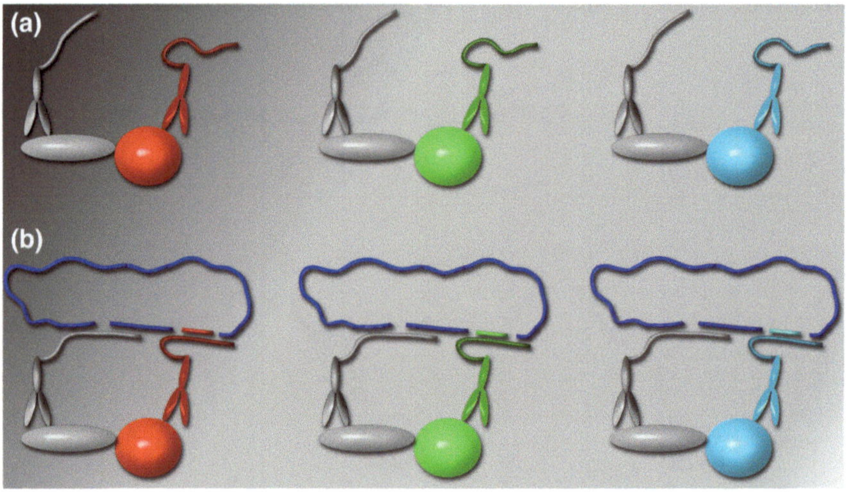


**Fig. 8** Multiplexed in situ PLA read-out. **a** The PLA probes are equipped with unique oligonucleotides as identifiers for their protein targets. In a situation with alternative interaction partners (indicated by *red*, *green*, and *cyan*) the corresponding PLA probes will bind in proximity to the PLA probe against the common interaction partner (*gray*). **b** The same circularization oligonucleotides (*blue*) hybridize to all probe combinations. The sequences of the PLA probes differ only in the central region where a tag sequence has been introduced. The tag sequence is subsequently used to label the RCA product with fluorophore-labeled detection oligonucleotides, generating differently colored RCA products representative of the different interaction events.

by a PLA probe, the enzymatic steps will not be 100 % efficient, and non-circular ligation products may be formed.

## 4 Read-Out Platforms

Detection of the PLA signal (i.e., RCA products) can be achieved using several different read-out method development. The very bright fluorophore-labeled RCA products that are produced in the in situ PLA are readily detectable over background using manual fluorescence microscopes or fluorescence microscopy scanners for high-throughput analyses (Leuchowius et al. 2010). In order to make the enumeration of all PLA signals in a tissue section or cell possible, it is essential that the signals are in focus. Therefore higher magnification images must be collected at several focal planes. Other read-out platforms that utilize fluorescence, such as flow cytometers (Leuchowius et al. 2009), enable high-throughput analysis of cells in suspension. As flow cytometry does not produce images, individual RCA products are not identified; instead total fluorescence intensity per cell is recorded. If the analyzed protein, protein complex, or PTM does not have to be detected in situ, PLA can be used to detect proteins bound to solid supports. In situ PLA has been used as a read-out for western blot and provides advantages of both

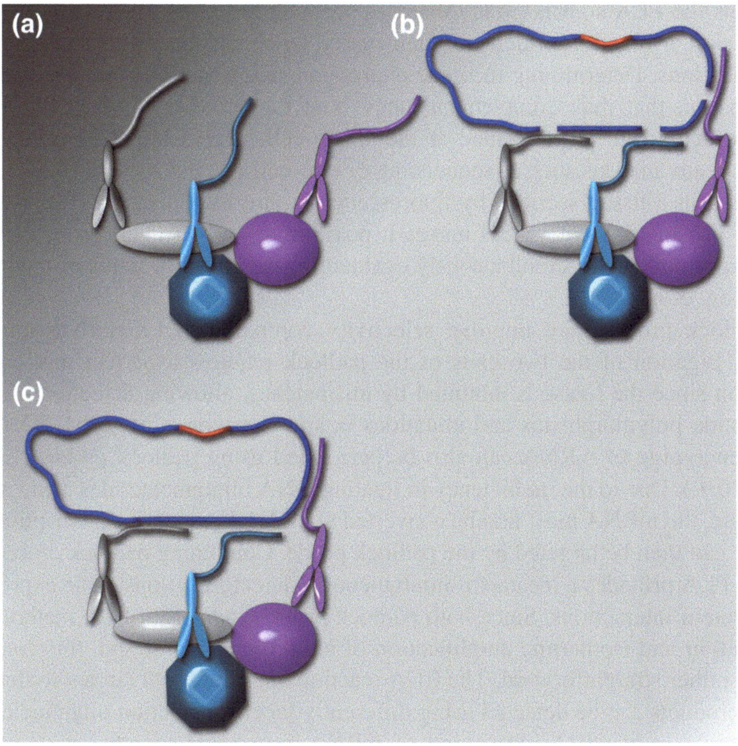


**Fig. 9** Detection of complexes consisting of three interacting proteins using in situ PLA. **a** The three PLA probes bind to their target proteins. **b** Hybridization of all three circularization oligonucleotides are needed to generate a circular DNA molecule. **c** Ligation of the gaps generates a circle that can be amplified using RCA

selectivity (since two antibodies have to agree on the identity of the proteins present in a given band on the blot) and sensitivity (since the signal can be amplified through RCA) (Liu et al. 2011).

If colorimetric read-out is preferred, the detection oligonucleotides can be equipped with enzymes such as horseradish peroxidase or alkaline phosphatase rather than fluorophores (Zieba et al. 2010). Alternatively, the fluorophore-labeled RCA products can be stained with enzyme-labeled anti-fluorophore antibodies (Paulsson et al. 2011). Other possible reporters may be conjugated to the detection oligonucleotides; for example, radioactive isotopes or gold particles may be used for electron microscopy.

## 5 Combining PLA with Other Method Development

In situ PLA can also be combined with other method development to provide additional information. The combination of immunofluorescence (IF) and in situ PLA may be used to determine the cell type or subcellular compartment that contains the PLA signals. Many diseases are manifested by alterations in protein expression or function, which are often caused by genetic aberrations or epigenetic modifications. Determining the gene expression status can thus provide important information that may complement analysis at the proteomic level. Analysis of nucleic acids can be performed at the single-cell level using PCR followed by downstream analysis such as sequencing or microarrays, but it can also be directly analyzed in a tissue sections by fluorescence in situ hybridization (FISH). Combining in situ PLA with FISH makes it possible to look at interactions of deregulated proteins while simultaneously evaluating amplification status of their genes (Renfrow et al. 2011).

Padlock probes may improve selectivity compared to FISH (Nilsson et al. 1994). Ligation of the two ends of the padlock requires a perfect match at the junction since the ligase is inhibited by mismatches, allowing detection of single nucleotide polymorphisms and mutations in situ (Larsson et al. 2004). Detection and genotyping of mRNA can also be performed using padlock probes (Larsson et al. 2010). Due to the inefficiency in ligating DNA oligonucleotides using a RNA template, the mRNA must first be converted to cDNA by reverse transcription. The cDNA can then be targeted by the padlock probe. Combining padlock probes with in situ PLA provides a means to simultaneously detect mutations, gene expression, and protein interactions. Since both padlock probes and in situ PLA methods rely on ligation and isothermic amplification of a circular DNA strand, this combination is rather straightforward. The RCA reaction amplifies both circles so that both RCA products can be detected using differently labeled detection oligonucleotides (Weibrecht et al. 2011). Detection of mRNA expression together with protein-nucleic acid interaction and PTMs is invaluable when mapping signaling pathway activity from receptor–ligand interactions through expression of the genes affected by the pathway.

## 6 Considerations

As with other antibody-based methods, in situ PLA requires high-quality affinity reagents to perform optimally. Furthermore, optimal conditions for antigen retrieval and blocking and antibody concentrations must be determined for each new assay. The requirement for dual recognition puts further constraints on the method since the antigenic epitopes have to be available for targeting and not obscured by the interaction surface in a protein interaction. Positive and negative controls, both technical and biological, are important during validation of the assays. The technical controls ensure that the assay is working and can include PLA probes against different targets, using blocking peptides, and omitting enzymes. To ensure that the PLA probes do bind the cells immuno-RCA can be performed using a padlock probe targeting the oligonucleotide arm of the PLA probe. Biological controls such as cells that do and do not express the proteins of interest can be used to validate the selectivity of the PLA probes. Controls should also include conditions that induce or block interactions of interest and may include drugs or transfection of fusion constructs of the interaction partners or a competitor of the interaction (Baan et al. 2010). As further reading, we recommend a few detailed protocols for different in situ PLA applications with extensive troubleshooting advice (Soderberg et al. 2008; Leuchowius et al. 2011; Weibrecht et al. 2013).

# 7 Future Perspectives

Herein, we have described how the in situ PLA method has evolved since its development in 2006 to take advantage of the vast repertoire of enzymes that modify DNA and the strict rules of DNA base pairing. We expect to see several modifications of the method in the near future that further expand its usability. The ability to visualize protein interactions and modifications in situ, together with the potential for high multiplexing are key features of in situ PLA that make it a powerful tool for analysis of functional status of proteins (e.g., analysis of signaling networks). The major advantage of the method is its potent signal amplification, resulting from the RCA, that enables detection of single molecules in tissue sections. The diagnostic potential of in situ PLA has not yet been thoroughly evaluated, but in situ PLA analysis of PTK6-HER2 interactions has prognostic relevance (Aubele et al. 2010). In situ PLA has been used to study the consequences of mutations in EGFR with respect to phosphorylation (Chen et al. 2013), response to EGFR inhibitors (Gajadhar et al. 2012), and how E-cadherin interactions with its binding partners are affected by E-cadherin missense mutations (Figueiredo et al. 2013) and P-cadherin expression (Ribeiro et al. 2013) .

In situ PLA is not limited to detection of interactions between proteins. It can also be used to detect interactions of nucleic acids and proteins. Detection of protein-DNA interactions is hampered by the need to render the genomic DNA single stranded. In recent years exciting new technologies have been developed that allow detection of double-stranded DNA using artificial proteins such as artificial zinc finger proteins (ZFPs). ZFPs are modular: These synthetic polypeptides have three to six recognition domains that each bind to the major groove of double-stranded DNA and sequence-specifically recognize three base pairs. A ZFP with six recognition domains thus binds to an 18-bp stretch of double-stranded DNA in a sequence-specific manner. Coupling ZFPs to oligonucleotides and using these constructs as proximity probes might be an interesting approach to directly target double-stranded DNA for analysis of protein-DNA interactions. This might allow the detection of transcription factor binding and thus will open a whole new world of in situ analyses.

The next challenge for in situ PLA—and where the method might provide the largest impact—is to use a highly multiplexed assay that in parallel targets several nodes in several signaling networks. By using such an assay we will be in a position to comprehensively study cellular communication in tumors. We could obtain information on what networks are active in cancer cells, and to what extent this varies depending on positioning within the tumor. This strategy may also reveal how the surrounding non-malignant cells in the tumor microenvironment interact with the cancer cells. This knowledge will enable better diagnostics, improved prediction of response to therapy, and possibly also aid the development of novel drugs that can modify the microenvironment to reduce cancer growth and ability to metastasize.

## References

Aubele M, Spears M, Ludyga N, Braselmann H, Feuchtinger A, Taylor KJ, Lindner K, Auer G, Stering K, Hofler H, Schmitt M, Bartlett JM (2010) In situ quantification of HER2-protein tyrosine kinase 6 (PTK6) protein–protein complexes in paraffin sections from breast cancer tissues. Br J Cancer 103:663–667

Baan B, Pardali E, ten Dijke P, van Dam H (2010) In situ proximity ligation detection of c-Jun/AP-1 dimers reveals increased levels of c-Jun/Fra1 complexes in aggressive breast cancer cell lines in vitro and in vivo. Mol Cell Proteomics 9:1982–1990

Burke DH, Gold L (1997) RNA aptamers to the adenosine moiety of S-adenosyl methionine: structural inferences from variations on a theme and the reproducibility of SELEX. Nucleic Acids Res 25:2020–2024

Carpenter AE, Jones TR, Lamprecht MR, Clarke C, Kang IH, Friman O, Guertin DA, Chang JH, Lindquist RA, Moffat J, Golland P, Sabatini DM (2006) Cell profiler: image analysis software for identifying and quantifying cell phenotypes. Genome Biol 7:R100

Chen TC, Liu YW, Huang YH, Yeh YC, Chou TY, Wu YC, Wu CC, Chen YR, Cheng HC, Lu PJ, Lai JM, Huang CY (2013) Protein phosphorylation profiling using an in situ proximity ligation assay: phosphorylation of AURKA-elicited EGFR-Thr654 and EGFR-Ser1046 in lung cancer cells. PLoS One 8:e55657

Clausson CM, Allalou A, Weibrecht I, Mahmoudi S, Farnebo M, Landegren U, Wahlby C, Soderberg O (2011) Increasing the dynamic range of in situ PLA. Nat Methods 8:892–893

Conze T, Carvalho AS, Landegren U, Almeida R, Reis CA, David L, Soderberg O (2010) MUC2 mucin is a major carrier of the cancer-associated sialyl-Tn antigen in intestinal metaplasia and gastric carcinomas. Glycobiology 20:199–206

Coons AH, Creech HJ, Jones RN, Berliner E (1942) The demonstration of pneumococcal antigen in tissues by the use of fluorescent antibody. J Immunol 45:159–170

Figueiredo J, Soderberg O, Simoes-Correia J, Grannas K, Suriano G, Seruca R (2013) The importance of E-cadherin binding partners to evaluate the pathogenicity of E-cadherin missense mutations associated to HDGC. Eur J Hum Genet 21:301–309

Fredriksson S, Gullberg M, Jarvius J, Olsson C, Pietras K, Gustafsdottir SM, Ostman A, Landegren U (2002) Protein detection using proximity-dependent DNA ligation assays. Nat Biotechnol 20:473–477

Gajadhar A, Guha A (2010) A proximity ligation assay using transiently transfected, epitope-tagged proteins: application for in situ detection of dimerized receptor tyrosine kinases. Biotechniques 48:145–152

Gajadhar AS, Bogdanovic E, Munoz DM, Guha A (2012) In situ analysis of mutant EGFRs prevalent in glioblastoma multiforme reveals aberrant dimerization, activation, and differential response to anti-EGFR targeted therapy. Mol Cancer Res 10:428–440

Gomez D, Shankman LS, Nguyen AT, Owens GK (2013) Detection of histone modifications at specific gene loci in single cells in histological sections. Nat Methods 10:171–177

Goransson J, Wahlby C, Isaksson M, Howell WM, Jarvius J, Nilsson M (2009) A single molecule array for digital targeted molecular analyses. Nucleic Acids Res 37:e7

Gu GJ, Friedman M, Jost C, Johnsson K, Kamali-Moghaddam M, Pluckthun A, Landegren U, Soderberg O (2013) Protein tag-mediated conjugation of oligonucleotides to recombinant affinity binders for proximity ligation. N Biotechnol 30:144–152

Gullberg M, Gustafsdottir SM, Schallmeiner E, Jarvius J, Bjarnegard M, Betsholtz C, Landegren U, Fredriksson S (2004) Cytokine detection by antibody-based proximity ligation. Proc Natl Acad Sci U S A 101:8420–8424

Gustafsdottir SM, Schlingemann J, Rada-Iglesias A, Schallmeiner E, Kamali-Moghaddam M, Wadelius C, Landegren U (2007) In vitro analysis of DNA-protein interactions by proximity ligation. Proc Natl Acad Sci U S A 104:3067–3072

Jarvius M, Paulsson J, Weibrecht I, Leuchowius KJ, Andersson AC, Wahlby C, Gullberg M, Botling J, Sjoblom T, Markova B, Ostman A, Landegren U, Soderberg O (2007) In situ detection of phosphorylated platelet-derived growth factor receptor beta using a generalized proximity ligation method. Mol Cell Proteomics 6:1500–1509

Jung J, Lifland AW, Zurla C, Alonas EJ, Santangelo PJ (2013). Quantifying RNA-protein interactions in situ using modified-MTRIPs and proximity ligation. Nucleic Acids Res 41: e12

Koos B, Paulsson J, Jarvius M, Sanchez BC, Wrede B, Mertsch S, Jeibmann A, Kruse A, Peters O, Wolff JE, Galla HJ, Soderberg O, Paulus W, Ostman A, Hasselblatt M (2009) Platelet-derived growth factor receptor expression and activation in choroid plexus tumors. Am J Pathol 175:1631–1637

Larsson C, Grundberg I, Soderberg O, Nilsson M (2010) In situ detection and genotyping of individual mRNA molecules. Nat Methods 7:395–397

Larsson C, Koch J, Nygren A, Janssen G, Raap AK, Landegren U, Nilsson M (2004) In situ genotyping individual DNA molecules by target-primed rolling-circle amplification of padlock probes. Nat Methods 1:227–232

Leuchowius KJ, Clausson CM, Grannas K, Erbilgin Y, Botling J, Zieba A, Landegren U (2013) Söderberg O [2]Parallel visualization of multiple protein complexes in individual cells in tumor tissue[2]. Mol Cell Proteomics 12(6):1563–1571

Leuchowius KJ, Jarvius M, Wickstrom M, Rickardson L, Landegren U, Larsson R, Soderberg O, Fryknas M, Jarvius J (2010) High content screening for inhibitors of protein interactions and post-translational modifications in primary cells by proximity ligation. Mol Cell Proteomics 9:178–183

Leuchowius KJ, Weibrecht I, Landegren U, Gedda L, Soderberg O (2009) Flow cytometric in situ proximity ligation analyses of protein interactions and post-translational modification of the epidermal growth factor receptor family. Cytometry A 75:833–839

Leuchowius KJ, Weibrecht I, Soderberg O (2011) In situ proximity ligation assay for microscopy and flow cytometry. Curr Protoc Cytom. doi:10.1002/0471142956.cy0936s56

Liu Y, Gu J, Hagner-McWhirter A, Sathiyanarayanan P, Gullberg M, Soderberg O, Johansson J, Hammond M, Ivansson D and Landegren U (2011). Western blotting via proximity ligation for high performance protein analysis. Mol Cell Proteomics 10: O111 011031

Matic I, Schimmel J, Hendriks IA, van Santen MA, van de Rijke F, van Dam H, Gnad F, Mann M, Vertegaal AC (2010) Site-specific identification of SUMO-2 targets in cells reveals an inverted SUMOylation motif and a hydrophobic cluster SUMOylation motif. Mol Cell 39:641–652

Mocanu MM, Varadi T, Szollosi J, Nagy P (2011) Comparative analysis of fluorescence resonance energy transfer (FRET) and proximity ligation assay (PLA). Proteomics 11:2063–2070

Nilsson M, Malmgren H, Samiotaki M, Kwiatkowski M, Chowdhary BP, Landegren U (1994) Padlock probes: circularizing oligonucleotides for localized DNA detection. Science 265:2085–2088

Paulsson J, Lindh MB, Jarvius M, Puputti M, Nister M, Nupponen NN, Paulus W, Soderberg O, Dresemann G, von Deimling A, Joensuu H, Ostman A, Hasselblatt M (2011) Prognostic but not predictive role of platelet-derived growth factor receptors in patients with recurrent glioblastoma. Int J Cancer 128:1981–1988

Pinto R, Carvalho AS, Conze T, Magalhaes A, Picco G, Burchell JM, Taylor-Papadimitriou J, Reis CA, Almeida R, Mandel U, Clausen H, Soderberg O, David L (2012) Identification of new cancer biomarkers based on aberrant mucin glycoforms by in situ proximity ligation. J Cell Mol Med 16:1474–1484

Renfrow JJ, Scheck AC, Dhawan NS, Lukac PJ, Vogel H, Chandler JP, Raizer JJ, Harsh GR, Chakravarti A, Bredel M (2011) Gene-protein correlation in single cells. Neuro Oncol 13:880–885

Ribeiro AS, Sousa B, Carreto L, Mendes N, Nobre AR, Ricardo S, Albergaria A, Cameselle-Teijeiro JF, Gerhard R, Soderberg O, Seruca R, Santos MA, Schmitt F, Paredes J (2013) P-cadherin functional role is dependent on E-cadherin cellular context: a proof of concept using the breast cancer model. J Pathol 229:705–718

Schubert W, Bonnekoh B, Pommer AJ, Philipsen L, Bockelmann R, Malykh Y, Gollnick H, Friedenberger M, Bode M, Dress AW (2006) Analyzing proteome topology and function by automated multidimensional fluorescence microscopy. Nat Biotechnol 24:1270–1278

Sehat B, Tofigh A, Lin Y, Trocme E, Liljedahl U, Lagergren J and Larsson O (2010). SUMOylation mediates the nuclear translocation and signaling of the IGF-1 receptor. Sci Signal 3: ra10

Soderberg O, Gullberg M, Jarvius M, Ridderstrale K, Leuchowius KJ, Jarvius J, Wester K, Hydbring P, Bahram F, Larsson LG, Landegren U (2006) Direct observation of individual endogenous protein complexes in situ by proximity ligation. Nat Methods 3:995–1000

Soderberg O, Leuchowius KJ, Gullberg M, Jarvius M, Weibrecht I, Larsson LG, Landegren U (2008) Characterizing proteins and their interactions in cells and tissues using the in situ proximity ligation assay. Methods 45:227–232

Tavoosidana G, Ronquist G, Darmanis S, Yan J, Carlsson L, Wu D, Conze T, Ek P, Semjonow A, Eltze E, Larsson A, Landegren UD, Kamali-Moghaddam M (2011) Multiple recognition assay reveals prostasomes as promising plasma biomarkers for prostate cancer. Proc Natl Acad Sci U S A 108:8809–8814

Tuerk C, Gold L (1990) Systematic evolution of ligands by exponential enrichment: RNA ligands to bacteriophage T4 DNA polymerase. Science 249:505–510

Weibrecht I, Gavrilovic M, Lindbom L, Landegren U, Wahlby C, Soderberg O (2012) Visualising individual sequence-specific protein-DNA interactions in situ. N Biotechnol 29:589–598

Weibrecht I, Grundberg I, Nilsson M, Soderberg O (2011) Simultaneous visualization of both signaling cascade activity and end-point gene expression in single cells. PLoS One 6:e20148

Weibrecht I, Leuchowius KJ, Clausson CM, Conze T, Jarvius M, Howell WM, Kamali-Moghaddam M, Soderberg O (2010) Proximity ligation assays: a recent addition to the proteomics toolbox. Expert Rev Proteomics 7:401–409

Weibrecht I, Lundin E, Kiflemariam S, Mignardi M, Grundberg I, Larsson C, Koos B, Nilsson M, Soderberg O (2013) In situ detection of individual mRNA molecules and protein complexes or post-translational modifications using padlock probes combined with the in situ proximity ligation assay. Nat Protoc 8:355–372

Zieba A, Pardali K, Soderberg O, Lindbom L, Nystrom E, Moustakas A, Heldin CH and Landegren U (2012). Intercellular variation in signaling through the TGF-beta pathway and its relation to cell density and cell cycle phase. Mol Cell Proteomics 11: M111 013482

Zieba A, Wahlby C, Hjelm F, Jordan L, Berg J, Landegren U, Pardali K (2010) Bright-field microscopy visualization of proteins and protein complexes by in situ proximity ligation with peroxidase detection. Clin Chem 56:99–110

# Cytobank: Providing an Analytics Platform for Community Cytometry Data Analysis and Collaboration

**Tiffany J. Chen and Nikesh Kotecha**

**Abstract** Cytometry is used extensively in clinical and laboratory settings to diagnose and track cell subsets in blood and tissue. High-throughput, single-cell approaches leveraging cytometry are developed and applied in the computational and systems biology communities by researchers, who seek to improve the diagnosis of human diseases, map the structures of cell signaling networks, and identify new cell types. Data analysis and management present a bottleneck in the flow of knowledge from bench to clinic. Multi-parameter flow and mass cytometry enable identification of signaling profiles of patient cell samples. Currently, this process is manual, requiring hours of work to summarize multi-dimensional data and translate these data for input into other analysis programs. In addition, the increase in the number and size of collaborative cytometry studies as well as the computational complexity of analytical tools require the ability to assemble sufficient and appropriately configured computing capacity on demand. There is a critical need for platforms that can be used by both clinical and basic researchers who routinely rely on cytometry. Recent advances provide a unique opportunity to facilitate collaboration and analysis and management of cytometry data. Specifically, advances in cloud computing and virtualization are enabling efficient use of large computing resources for analysis and backup. An example is Cytobank, a platform that allows researchers to annotate, analyze, and share results along with the underlying single-cell data.

T. J. Chen · N. Kotecha (✉)
Cytobank, Inc, Mountain View, CA, USA
e-mail: nikesh@cytobank.org

T. J. Chen
e-mail: tjchen@cytobank.org

Current Topics in Microbiology and Immunology (2014) 377: 127–157 
DOI: 10.1007/82_2014_364

Published Online: 5 March 2014

## Contents



# 1 Introduction

Cytometry is evolving as a field. The ability to use intra and extracellular properties to quantify cell type and cell signal and cell cycle events has the potential for tremendous benefits in diagnosis and drug discovery. As a result, instrument technologies such as the CyTOF (DVS Sciences) have emerged to simultaneously measure 40 parameters (and theoretically up to 100) (Bandura et al. 2009). This new approach termed mass cytometry allows scientists and clinicians to ask questions that were previously not possible to address with single cell assays (the largest number of markers measured with fluorescence cytometry to date is 20). Other methods such as single-cell gene expression analysis and new computational methods for extracting cell-specific information from gene expression data (Shen-Orr et al. 2010) are still

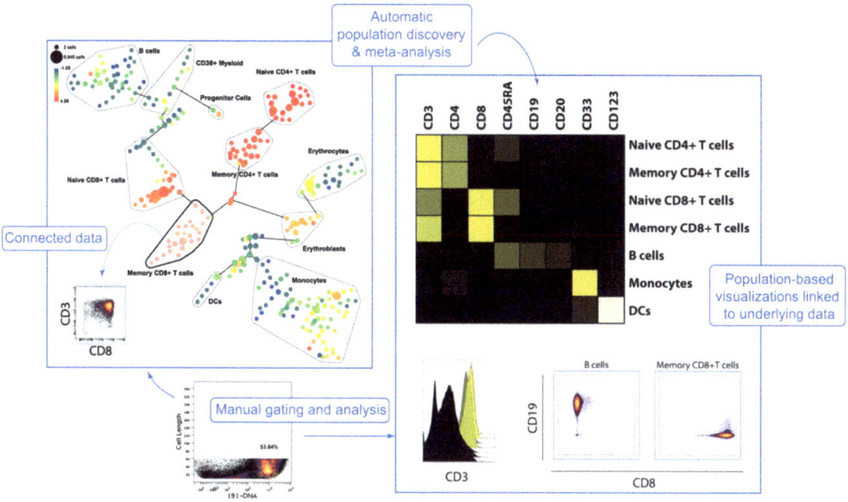


**Fig. 1** Achieving expert-driven biological results by leveraging high-dimensional analysis methods. Cytometry analysis primarily focuses on the discovery of expert-derived biological populations. This figure shows two different methods for identification of these populations, SPADE bubbling and manual gating. The subsequent meta-analysis of these populations involves heatmaps, histogram overlays, and 2D plots. Being able to move back and forth between population discovery and these high-level figures are not only essential for comprehension of the biological results but also for effective results, communication, and reproducible research

considered approximations when compared to the ability to directly measure proteomics events (e.g., phosphorylation of pStat5) in single cells via cytometry. The possibilities opened by mass cytometry and the growing interest in systems level immunology studies position cytometry to be a key player in biological and medical research in the years to come.

These breakthroughs have underscored the need for platforms that can scale to handle the analysis and storage needs required to enable the science. Current analysis solutions are designed to handle typical flow cytometry experiments, where the key need until now has been focused on the manual process of gating—the identification of cells of interest—by visual inspection. In experiments that can easily involve hundreds of samples and include simultaneous measurement of 40 or more parameters, these solutions do not scale. There is a need for novel algorithms and visualizations to identify populations of interest. The interdisciplinary nature of analyzing this information requires a platform enabling novel analyses and visualizations where users can organize and manage their experiments, communicate results with their collaborators, have annotations linked to analyses, and provide a way to scale compute resources as needed.

For example, consider Fig. 1, which shows a result from a typical high-dimensional CyTOF experiment. The workflow involved various levels of analyses including conventional processing using manual gating, clustering approaches such as Spanning-tree Progression Analysis of Density-normalized Events (SPADE) to group similar cells, and heatmap overviews to summarize various pieces of information. This example result is focused on only one sample, exposed to one condition in an experiment. It is relatively straightforward to design and generate experiments where tens of populations are identified and compared across hundreds of samples comprising many conditions, doses, and individuals. How does one work with this dataset? How does one effectively communicate these results? How can collaborations across disciplines and often across geographies be facilitated? How does one do this securely and in a clinical setting?

This chapter addresses these issues. In Sect. 2, we will define the challenges involved in handling cytometry data. In Sect. 3, we will highlight the approaches taken by the Cytobank platform to address these challenges. In Sect. 4, we will talk about novel approaches and next steps in the cytometry analysis process. Section 5 will walk the readers through the specific steps that lead to generation of the panels shown in Fig. 1. We will conclude with the future directions in Sect. 6.

## 2 Building a Platform for Cytometry Data Analysis and Collaboration

The breakthroughs in high-throughput, single cell cytometry have created challenges in the storage, analysis, and representation of data. It is not uncommon for a laboratory to generate hundreds of megabytes to gigabytes of data in a single cytometry experiment. Core facilities have an even higher throughput. These records contain data on millions of cells and tens of measurements per cell.

### *2.1 Current Cytometry Workflow is Fragmented*

Cytometry sample preparation, data collection, and analysis occur in a series of stages where many individuals play different roles. These include but are not limited to:

1. Sample acquisition
2. Experiment preparation and execution
3. Collection of data from a cytometer
4. Preprocessing, quality assurance, and quality control of the data
5. Qualitative data analysis
6. Quantitative data analysis and visualization of relevant metrics
7. Figure generation and results presentation.

Individual users, overwhelmed with data complexity, often cope by simplifying single cell data into aggregate statistics, thus limiting potential insights that can be gleaned from these data. Gaining richer insight into human disease using this technology is rarely limited to one individual. Instead, experiments of this magnitude often require multidisciplinary teams that include clinicians, molecular biologists, statisticians, and informaticians who understand biology and medicine, and have shared but also distinct goals at each stage. More commonly, achieving these goals requires individual researchers working collectively to achieve a biological result.

The number of data processing steps, coupled with the multiple researchers in the analysis pipeline, results in fragmented data analysis. Current workflows usually involve the use of multiple applications resulting in fragmented output. As results are often shared as processed figures, raw data are disjointed from the aggregate summaries making re-evaluation of the original data difficult. This also hinders the reproducibility of research. Facilitating communication among researchers is crucial, as the cytometry field is still maturing, and methods for data preprocessing (compensations, transformations) are not standardized (standard ways of annotating and analyzing data are under discussion but not fully agreed upon) (Tung et al. 2004; Bagwell 2005; Parks et al. 2006).

In the typical workflow, users first outline their experiment thoughts and design, including proteins and compounds of interest in their lab notebook. Data are collected using software provided by the flow cytometry vendor, which outputs a file in the flow cytometry standard (FCS) format (Spidlen et al. 2010; Qiu et al. 2011). In addition to raw data, the FCS specifications provide areas for incorporating metadata and analysis components using keyword/value pairs along with the raw data. These components, if used, are not generally recorded in a standardized fashion or in sufficient detail to use by independent parties. Prior to analysis, users typically perform two preprocessing steps. The first is compensation, which is the correction of spectral overlap between two fluorophores. The second is transformation of the data into a log or log-like space. With new technologies such as mass cytometry, techniques including file concatenation, normalization, and debarcoding are also added to this list, whereas others such as compensation are removed. After compensation and transformation, users will begin analysis by specifying a live cell gate and then the population of interest based on the presence/absence of particular markers. The visualization of choice for display during this process is commonly depicted through histograms or contour plots. The procedure used to identify or gate cells is manual and (potentially) subjective but has been in common use since the advent of cytometry. Users prefer to follow a hierarchical path of 2-D scatter or contour plots during the subset identification process. After populations of interest are identified, levels are compared. Methods for doing this vary between labs and can often lead to debates about biological conclusions, since the results can be affected by the preprocessing steps above and are subject to human interpretation of the experimental controls. A method that has worked well is calculating the median response value for each marker and converting it into a fold change by comparing it to a reference sample (Irish et al. 2004). This statistic is often displayed in a heatmap format and often used as an endpoint to communicate the results of the experiment.

### 2.2 Analysis Workflow Requirements for Translational and Systems Level Research Using Cytometry Data Must Involve Collaborative Data Management

Experimental design is a key part of any research project. Often, many individuals interact during the design and execution of a flow cytometry experiment. Roles include, but are not limited to the following, some of which may overlap:

- Primary researcher (often the designer of the experiment)
- Primary investigator (the lead researcher for an experiment or greater project)
- Computational biologist and/or statistician
- Flow cytometry core researcher/technician
- Clinical investigator.

In order to execute a cytometry experiment in an era of large-scale translational research, teams of individuals aggregate this design over presentation slide decks, lab notebooks, and documented protocols. At the more immediate level, other information includes the antibodies and reagents used, as well as technical and biological controls.

*There is a need for centralized data sharing.* Analysis workflow requirements for translational and systems level research using cytometry data must involve collaborative data management. The expansion of cytometry to measure patient-specific signaling profiles has positioned it as a key player in clinical translational research and for systems-wide studies. Its potential for adoption is large due to the large number of cytometers already placed in the clinic. Breaks in the analysis pipeline can be attributed to a few problems, one of which is centralized data sharing. First, it is currently difficult for biological researchers to access or query clinically annotated cytometry data. Second, these researchers do not always leverage newer data analysis and novel machine learning methods that take advantage of the single-cell information (discussed in other chapters), since it is rare for these methods to be directly accessed by the experimental flow cytometry community. Cross-discipline analysis requires a balancing of technical detail necessary to convey a result or a potential problem across researchers. Clinicians and biologists often want a digested version of the results so they can focus on understanding the disease at hand, whereas statisticians need access to quality-controlled, annotated raw data and methods used in preprocessing steps.

### 2.3 Communication of Metadata: Effective Annotation is Required for Sharing of Cytometry Datasets

FCS files are not independent entities. External information about an experiment is essential to the biological interpretation of the data in an FCS file. Users typically have this metadata in the form of protocols, spreadsheets, and sample annotations in lab notebooks that are not required entries during data acquisition from a cytometer. This information can include proteins or biomarkers being measured (e.g., phoshpo-STAT5), the conditions (e.g., stimulation with GM-CSF), the cell types of interest (e.g., B cells), the model organism (e.g., Balb/C mice), the disease of interest (e.g., leukemia), and protocol information (e.g., method of permeabilization and staining volume). A list of metadata related to sample preparation and data acquisition are listed (but not limited to) below:

- Proteins/biomarkers and the channels on which relevant data are measured
- Conditions (e.g., cytokines, therapeutics)
- Time points and doses
- Reference IDs to sample acquisition (e.g., de-identified IDs from patient data)
- Disease or cell type studied
- Plate annotations (columns, rows, plate identifiers).

Standardized ways of preserving this information throughout analysis is not only useful for communicating biological results through a collaborative workflow but is essential in large-scale studies where multiple experiments are collated. Standard nomenclature for this metadata requires incorporation of many terms, including:

- Protein identifiers [e.g., UniProt IDs (UniProt 2010), BRENDA (Gremse et al. 2011)]
- Taxonomy identifiers [e.g., NCBI taxonomy database (Wheeler et al. 2008)]
- Descriptions of cell types and cell sources [e.g., OBO-cell type (Smith et al. 2007)]
- Descriptions of diseases studied [e.g., ICD9 (CDC/National Center for Health Statistics 2009)]
- Cytometry file-specific information [e.g., MIFlowCyt (Lee et al. 2008)].

Although standardization of experiment metadata and use of controlled vocabularies are necessary, they face a tough adoption curve by the cytometry community. It is critical to determine incentives that will motivate biologists to annotate data and to provide a clear, helpful interface to ensure accuracy of annotations. Thus, the challenge is to make annotation appealing to the researcher as opposed to making annotation an annoyance or frustrating barrier. The growth in flow and mass cytometry datasets generated in a collaborative nature create an increased need for advanced statistical analyses, which depend heavily on the propagation of these annotations. Ideally, information for a group of samples (e.g., a patient cohort) would be securely available online and the data would be interconnected to allow for dynamic visualizations of the primary data along with the underlying analysis at the single-cell level.

## 2.4 *The Importance of Layered Analysis and Key Components*

Greater accessibility to raw data and a movement toward more reproducible research are paving the way for public data sharing and post-analysis of data. Research collaborations may span multiple institutions, but regardless of whether or not analysis has been completed at the individual, lab, or consortium level, there are challenges in review, publication, and post-publication analysis of the results from these complex datasets. Although scientific journals are able to display two-dimensional graphs and one-dimensional statistics, these figures necessarily leave out many layers of data that would be useful to other researchers in the field, either for verification of controls or for extracting novel conclusions from the experiments. In this section we introduce many of the key components in the analysis pipeline in addition to their interdependencies. Effectively communicating and distilling these analysis steps is essential to reproducible research, as well as to the expansion of flow cytometry to larger multi-parameter datasets in collaborative projects.

### 2.4.1 Data Preprocessing Affects the Downstream Analysis Pipeline

A number of preprocessing steps may be required for cytometry data analysis; the number varies depending on whether or not the sample is collected from a mass cytometer. Changes to some of these steps can also affect population finding and computational analyses. Storing preprocessing steps alongside cytometry data is an important but often ignored part of cytometry data analysis. Preprocessing of cytometry-based data can include file concatenation, data normalization, and de-barcoding. As discussed previously, prior to gating and SPADE analysis, other commonly used intermediate steps also include compensation (for fluorescence-based flow cytometry) and data transformation, which is the scaling of raw data along the axes. Data transformation is important for visualization and interpretation of the collected data during analysis, particularly during population finding.

### 2.4.2 Current Gold Standard in Cytometry Involves Manual Identification of Populations of Interest

Manual gating is a commonly used method for cytometry data analysis and is regarded as the gold standard for identifying populations. Gating involves identification of cell subsets by manually drawing geometric shapes around cell subsets. A key reason for this step in the workflow is that an expert can find a subpopulation of rare cells difficult to discover by automated methods. Manual gating, however, can vary across files and is difficult to reproduce across many experiments. This becomes especially difficult when comparing protein levels, instead of event counts, across populations.

### 2.4.3 Automated Methods to Identify Populations of Interest are Emerging

Automated methods are available for population finding, and this is an active area of research. Methods can be grouped into nonparametric and parametric statistical methods and include clustering based on k-means or expectation–maximization based mixture models (Lo et al. 2008; Pyne et al. 2009; Qian et al. 2010), spatial explorations of histograms (Roederer et al. 2001), variants combining the two (Roederer et al. 2001), and aggregation of cells based on density normalized events (Walther et al. 2009b; Qiu et al. 2011). The main avenue of deploying these algorithms is via packages for R/flowCore (Hahne et al. 2009), Matlab (Finck et al. 2013), Python (Frelinger et al. 2008), and Mathematica. Although these packages are widely used in the computational community and in cytometry research laboratories, it is rare that clinicians or biologists are facile with R or with scripting environments that are useful for parsing and managing flow cytometry data formats. Approaches such as GenePattern aim to make these tools available via interfaces but lack the data management and collaboration features also required for cross-disciplinary efforts (Spidlen et al. 2013). The informatics and statistical communities in cytometry face a key hurdle in making novel analyses and algorithms available to the broader cytometry community. Stand-alone algorithms are cumbersome. Users want to have results of the algorithm directly integrated with their primary data so they can pursue further analysis and share/communicate with their collaborators.

### 2.4.4 Layered Analysis Approaches that Connect Advanced Tools to Raw Data are Needed

The optimal software would incorporate data management, sharing, and advanced analyses under one roof. This approach can capture the (manual) process of identifying populations of interest and allow for layers of analysis to be built on top of stored raw data. Users could then focus on communicating their results and drill down to the technical details when exploring their data or incorporating new statistics. Another benefit of such an approach is extensibility of a centralized, layered analytic pipeline. Adding intermediate or end layers to analysis would no longer be a difficult task. One example of making advanced computational tools accessible to biologists is the incorporation of SPADE into Cytobank (Kotecha et al. 2010), layering it with the preprocessing steps prior to advanced analysis.

*A platform to centrally manage, analyze, and grow with the demand for cytometry is becoming a key requirement at large institutions.* There is a fast-growing need to add new computational algorithms as scalable layers to the analysis pipeline. Principal investigators and scientists in cytometry also recognize this hurdle. The throughput of cytometry data has increased both in the number of samples and in the ability to measure many features (e.g., mass cytometry assays), and users often find themselves exporting data to multiple applications for statistical tests or to access advanced algorithms and figure generation. A key hurdle is the inability to have results of the advanced algorithm directly integrated with primary data. In some labs, data-handling and reduction has become such a time-sink that investigators limit themselves to simpler analyses and miss opportunities to truly explore these rich datasets.

## 3 The Cytobank Platform and Approach

### *3.1 Cytobank has Emerged as a Platform to Allow Users to Manage, Share, and Analyze Cytometry Data Over the Web*

Recent advances in information technology have provided a unique opportunity to facilitate collaboration, data analysis, and management of cytometry data. Specifically, advances in cloud computing and virtualization are enabling ways to efficiently use large computing resources for analysis and backup via the Internet. Cytobank (www.cytobank.org) (Kotecha et al. 2010) is one approach and platform that builds on these advances to address challenges in working with flow and mass cytometry data. Much like using sites like Google, Facebook, or Amazon, with Cytobank, users do not have to think about computational capacity and can access or analyze their data from any web enabled device (e.g., laptop, iPad, smart phone).

Cytobank is the first platform for managing, sharing, and analyzing cytometry data that can be used over the web. Users can access, organize, analyze, and share their results with their group members. Data are stored centrally and securely and are accessible using a web browser to those with appropriate privileges. Cytobank allows visualization of common signaling profiles from a group of samples as well as analysis of the individual cell maps of signaling networks underlying the profile. Cytobank archives the original (raw) data and makes possible rich, extensible annotations of these data. It also provides the necessary framework for statistical analyses of the data with algorithms such as SPADE (Qiu et al. 2011; Linderman et al. 2012). The Cytobank platform has been the engine behind landmark papers in the emerging areas of high-dimensional and mass cytometry (Bendall et al. 2011; Bodenmiller et al. 2012).

### *3.2 Cytobank Motivates Data Annotation by Linking Experimental Details to Figure Creation*

Experiment annotations are usually written in a lab notebook and are not required during data collection. Thus, the data for each sample and condition are stored in binary flow cytometry standard (FCS) files, and there is often no explicit link between experimental information and the raw data. As a result, figures shown in publications are often far removed from the primary data, and annotation is perceived as a tedious task by experimenters. There is no apparent short-term benefit to annotation as the researcher would have to manually enter the information, and the researcher can refer to their lab notebook for experiment details during figure creation. Our approach in Cytobank is to use a flexible system of tags to capture information necessary for analysis. The annotations captured are in the form of "experiment variables" and can include controls, conditions (stimuli and inhibitors), sample types, individuals, time points, doses, replicates, and staining panels (typically listing antibody/dye combinations). Experiment variable tags are also used to group, order, and manipulate samples during figure creation; the quick, flexible system for creating figures motivates scientists to annotate data. Thus, there is a built-in reward system for users to encourage annotation—their flexibility in data analysis and figure creation increases as more information is ascribed to the experiment. Eventually, annotations will become part of experimental setup and not an afterthought.

### *3.3 Cytobank Links Aggregate Views to Underlying Single-Cell Data*

Fundamentally, cytometry analyses compare cell populations of interest. These comparisons often involve ordered sets of histograms or 2D contour plots. In Cytobank, the user arranges data in a layout using the experiment annotations and then views the data using customizable templates called "illustrations" (e.g., big histograms, small contour plots with outliers, a heatmap, overlaid histograms). With large datasets, a heatmap provides an overview of the data, but this view can distance users from the underlying single-cell data. As a compromise, the Cytobank allows users to toggle the view between the heatmap summary and that standard flow cytometry representation or to "view through" any square and see a plot of the data underlying it. Clicking a heatmap square brings a user to a detailed view of the data file used to calculate the statistic. This allows one to quickly jump from the overview to the raw data and to verify that the gates fit the data correctly or that the statistic accurately represents the underlying populations.

### *3.4 Cytobank can be Extended to Incorporate Novel Analyses and Visualizations for High-Dimensional Cytometry Data*

Statisticians and computational biologists working with flow cytometry data often require the ability to access data in statistical packages such as R. As a result, packages for analyzing flow cytometry data in R have been developed. Integration of the flowCore package (Hahne et al. 2009) and associated libraries into the popular computational biology framework known as BioConductor (Gentleman et al. 2004) is making flow cytometry accessible to a growing community of system biologists and bioinformaticians. Key hurdles still exist; one is the lack of associated information that makes the raw data in FCS files meaningful. These include gates used to identify cells of interest, markers measured in each channel of an FCS file, and experiment information such as sample ID, time point, and conditions (e.g., stimulus, drugs).

Biologists and clinicians are rarely facile with R. These researchers tend to export gated FCS files and an accompanying excel sheet with annotations to the statistician who in turn writes his/her own set of scripts to read the FCS files, ascribe the appropriate annotations, and run a clustering or advanced analysis. The statistician then returns a PDF or PowerPoint file with the analysis to the researcher. Key information such as how the cells of interest were identified/gated are ignored, and it is difficult to track potential errors due to mislabeling/re-processing of annotations. This loss of key knowledge is emerging as a fundamental problem as new technologies such as mass cytometry require new statistical and informatics approaches to handle 40+ simultaneous single cell measurements. Algorithms such as SPADE (Qiu et al. 2011; Linderman et al. 2012) have been developed to leverage the richness of such data but are difficult for biological researchers to use.

Cytobank provides a way for users to ascribe annotations as part of analysis and as a result capture the necessary information required for advanced statistics such as clustering. Currently, Cytobank can export statistics and other annotations which can then be imported into advanced statistical analysis programs. In the future, Cytobank will also enable the export of data and associated annotations from Cytobank into a user's R/flowCore. Technically savvy users can also choose to build on top of the Cytobank infrastructure to incorporate custom statistics and visualizations. Cytobank objects are designed using a layered approach to allow programmatic access at various levels. Programmers wishing to build on the raw data can access the files through the storage layer. The population management and gating interface allow access to compensated, gated, and transformed data.

## 3.5 *Cytobank Satisfies Data Security and Privacy Requirements for Large Institutions*

With research data, there are understandable concerns about data privacy and security at the user level and at the system level in the cloud. The Cytobank platform has been designed to satisfy requirements for security and privacy specified by large pharmaceutical companies and research institutions. Cytobank also satisfies guidelines from government agencies such as FISMA (NIST 2013) and has features that allow institutions to satisfy requirements for HIPAA and

HITECH (America 2009). Data on Cytobank servers are also encrypted and backed up off-site. In addition to system-level security, several layers of permissions are available at the user level. Data uploaded to Cytobank servers are visible only to the user until he or she chooses to share it with a collaborator. At that time, there are several options. Use of Cytobank Projects can give a collaborator view-only access to the results or can give a collaborator access to data for creation of new figures without permission to edit underlying annotations or gates. A user can also choose to give a collaborator full access to his or her experiment.

### 3.6 *Cytobank Reports Allow Effective Communication of Results*

Figures in a journal article represent the end-stage of the information spectrum. Though easy to read, static figures fail to capture details of an analysis and lack the depth of information necessary for true transparency and openness, particularly in a field like cytometry. The Cytobank platform allows for creation of web pages and sites that not only include plots and descriptions but also links to underlying analyses in Cytobank, gating hierarchy for the experiments shown, and sections for abstracts, protocols, and other pertinent details.

Cytobank Reports build on this by providing a public interface from which to view analyzed data and to access the raw data. Cytobank Reports meet existing flow cytometry data standards, including MIFlowCyt, created by the International Society for Analytical Cytology and American Chemical Society Analytical Cytometry Standards (Society 2013, Lee et al. 2008). These standards are also in consideration by other resources for publicly available cytometry data analysis, including Immport (Immport 2014) and Flow Repository (Spidlen et al. 2012). Both the size of the datasets (gigabytes) and the complexity of analyses in datasets generated by techniques such as mass cytometry preclude the utility of simple file-sharing (e.g., DropBox). With Cytobank Reports, plots and information can be organized neatly (much as they are in a journal article) and include links to directly access the analysis and the online data stored in Cytobank.

## 4 New Approaches and Next Steps in Cytometry Analysis

Thus far, we have reviewed the general strategies and needs for cytometry data analysis up through the individual sample level, with a minor emphasis on cross-sample comparisons. In reality, however, there are large growing classes of advanced cytometry analyses that include both complex biological and computational questions. These methods often utilize gating, but also build upon the general practice of population finding. Figure 2 is an example of how these extensions and new approaches are built on current analyses.

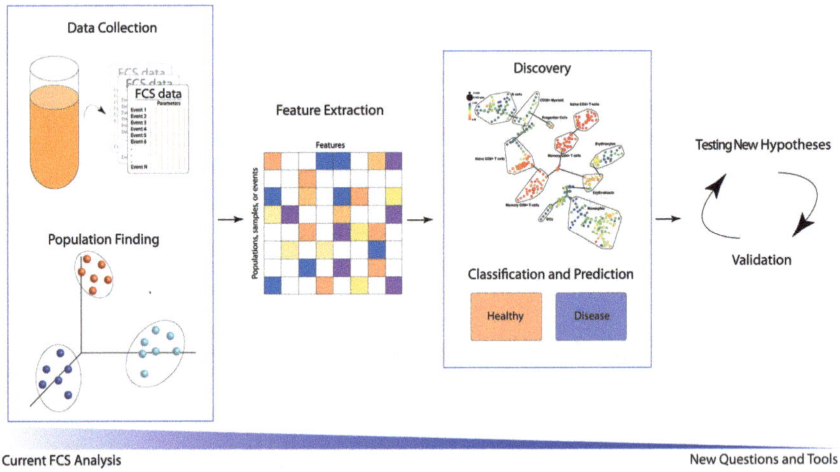


**Fig. 2** Next-generation workflow for cytometry data analysis. Through the incorporation of traditional and next-generation analysis methods, new testable hypotheses will be generated that can be validated either through experiments or literature review

After population finding, features are extracted that can then be used as the foundation for investigative analysis (labeled here as Discovery) as well as classification/outcomes prediction. From these results, achievable biological hypotheses are generated that are either tested through this process or validated through subsequent biological experimentation. In the case of computational results, researchers may validate their hypothesis by literature review. Overall, this process can be broken down into the following categories:

1. Feature extraction from flow cytometry data
2. Use of these features for:

a. Unsupervised exploration of the data (e.g., discovery of disease types)
b. Supervised investigation of the data (e.g., prediction of known disease types)

3. Hypothesis generation and biological validation (e.g., discovery of cell subsets that define a disease type, and the subsequent validation in a larger cohort).

One may look at this list, and ask: What does each stage mean biologically? For example, a large number of papers report population finding but ignore the later steps that lead to biological outcome analysis (Lo et al. 2008; Pyne et al. 2009; Walther et al. 2009a; Qian et al. 2010; Zare et al. 2010; Aghaeepour et al. 2011; Ge and Sealfon 2012). These stages can be most clearly explained by looking at papers that integrate these approaches. For example, dimensionality reduction techniques are often used in the discovery phase to differentiate between samples and disease progression (Bendall et al. 2011; Newell et al. 2012; Amir el et al. 2013). From the perspective of disease classification and prediction, post-clustering and features are typically used to differentiate groups with disease from healthy subjects. Research studies using well-established classifiers from the machine/statistical learning are only beginning to be performed (Bashashati et al. 2009; Zare et al. 2010; Aghaeepour et al. 2012). Use of classifiers for automated cytometry data analysis results in more robust hypothesis generation and enables the ability to rapidly reproduce and validate many of these findings either using flow cytometry data or through other experimental techniques. For those interested in learning more about cytometry bioinformatics in general, an excellent review is available (Wodak et al. 2013).

Post population-finding (either through manual gating or automated methods), visualization, and statistics surrounding sample analysis are crucial next steps. Fundamentally, cytometry analyses compare cell populations of interest. These across-sample analyses include, but are certainly not limited to:

- Visualizations:

- histograms
- dot plots colored by density
- contour plots with outliers
- network-based relationships

- Basic statistics:

- median/mean
- variance/standard deviation
- event counts
- percent of events in gate (for manual gating)
- correlation coefficient (Pearson)
- fold change.

Comparing samples both visually and statistically using these methods are relatively common within the cytometry community. It must be recognized, however, that nudging a gate or changing a transformation parameter can affect downstream steps. In Cytobank, the user arranges their data in a layout using the experiment, then annotates and views the data using customizable illustrations that are not dissociated from the underlying data. Cytobank is quickly emerging as a platform to house many of these new algorithms and visualization methods.

## 5 Example Analysis of a High-Dimensional CyTOF Experiment Using the Cytobank Platform

In the following section, we will illustrate a practical step-by-step analysis using the Cytobank platform. The starting data set is publicly available and thus users

**Fig. 3** An example of the data display post-upload in Cytobank. Note the three organizational categories: experiment files, compensation controls, and other controls

can perform each of these steps in their own web browser. For a more detailed explanation of how to perform a Cytobank analysis, readers of this chapter can also refer to a previous Cytobank publication (Kotecha et al. 2010).

## *5.1 Uploading Data and Creating a New Experiment*

Analysis in Cytobank begins with upload of cytometry data to the Cytobank platform. As an introduction to Cytobank analysis, we will use a publicly available CyTOF dataset (Bendall et al. 2011). After upload, raw files are displayed for the user to verify that valid files were uploaded, thus storing the original files prior to any post-processing layers. Figure 3 shows an example of the display of uploaded data.

Experiment Files, Compensation, and Other Controls are the three categories available for organization of FCS files prior to further analysis. Cytobank takes annotations from FCS filenames and autocategorizes these automatically. Files may be reassigned to other categories under the Channel pane setup in working illustration (see *Setting up Channels*). A click on "View Experiment Summary" will bring you to the experiment page, which contains high-level

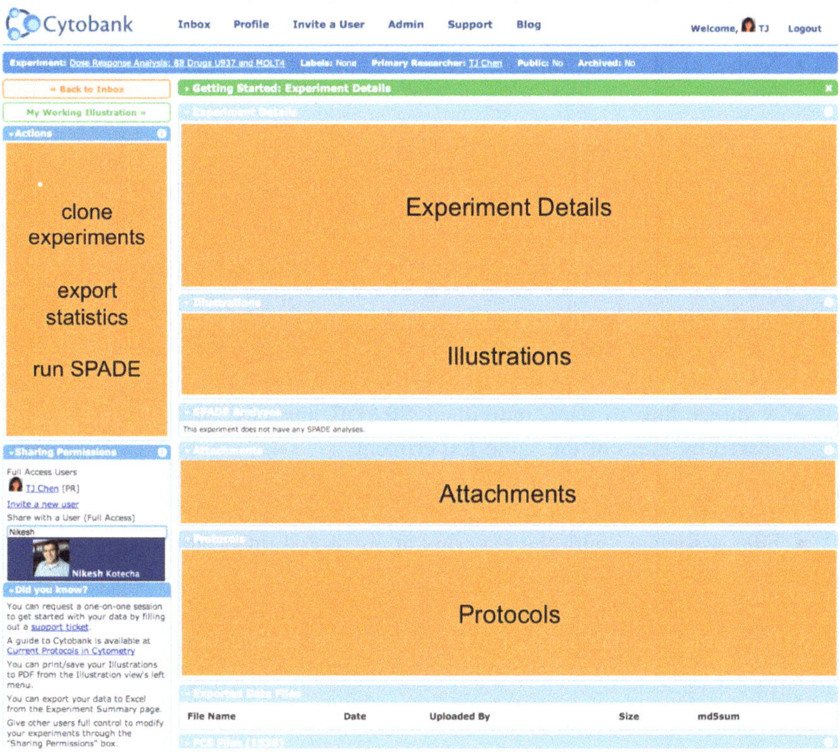


**Fig. 4** The experiment summary page for storage of raw FCS files, metadata, and annotations associated with an experiment, which also enables sharing of analyses with collaborators

information associated with the entire experiment. An example of the Experiment Summary page is shown in Fig. 4.

## 5.2 *Overview of the Experiment Details Page*

The experiment details page enables performance of high-level actions and use of tools relevant to the experiment. This page also links with metadata that is crucial to recreating both the experiment and the subsequent analyses of the cytometry data. This page includes:

*Actions*

- Includes the ability to clone an experiment (e.g., make a copy), export statistics, and run advanced tools such as SPADE.

*Experiment Details*

- Name of the experiment, names of Primary Researcher and Primary Investigator, and other information.

*Illustrations*

- One of the most powerful tools available, illustrations allow lay out of 2D plots, heatmaps, histogram overlays, and other figures, which are all connected to the underlying data. Snapshots of these can be generated as saved illustrations, which are static. A user's working illustration is dynamic and can be changed. All users share gates in an experiment.

*Spade Analyses*

- A direct link to each individual SPADE analysis. SPADE analyses are shared by users of the same Experiment.

*Attachments*

- Files associated with the experiment are stored under attachments.

*Protocols*

- Protocols associated with the experiment can be attached as files, URLs, or text directly written into Cytobank.

*Exported Data Files*

- Exported files, included statistics, are attached as zip files.

*Sharing Permissions*

- The entire experiment, including all analyses and annotations, can be shared with collaborators simply by typing in their names.

## 5.3 Setting up Channels, Conditions, and Populations

Under a new working illustration, the figure dimensions do not yet have information. Each individual FCS file contains a rich amount of information, including individual, condition, and multiple cell populations (Fig. 5).

### 5.3.1 Setting up Channels

In some cases, you may have multiple panels during staining. A click on the "Setup" link under Channels brings up the File Categorization and Panel Assignments page where files can be categorized as experimental, controls, or beads, new panels can be created, and reagents renamed. Renaming can be

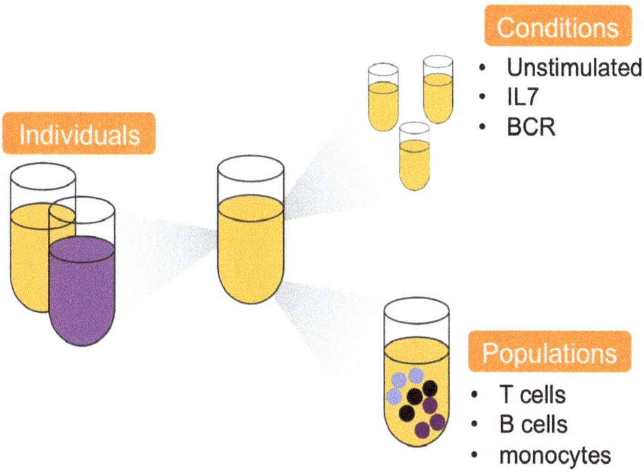


**Fig. 5** A few example dimensions annotated during the process of data collection or analysis

important when multiple FCS files have the same antigen measured on a common fluorophore or metal, but the reagent name was not matched during collection. Since this is a tutorial experiment, channels match between tubes, but this may not always be the case.

### 5.3.2 Setting up Conditions Automatically

A click on the small rectangle called Conditions toggles the dimension "on"—it will turn orange and a new box will appear next to the Channels and Populations figure dimensions. Select "Click to Setup" under the Conditions figure dimension. In this case, there are three conditions for this experiment, namely Basal (unstimulated), BCR, and IL7.

Type these three names into the box and then click "Add Conditions." You should see each FCS file become automatically categorized for the given conditions (Fig. 6). You can click and drag these files into other categories. In addition, you can click on the name of a condition to edit it. Note, naming of conditions and other dimension properties is case-sensitive.

### 5.3.3 Gating Populations

The next step is to gate set of intact cells (sometimes referred to as singlets depending on the channels used) and create a population (Fig. 7). Click on the "Gate" link in the Populations figure dimension. In this experiment, we will

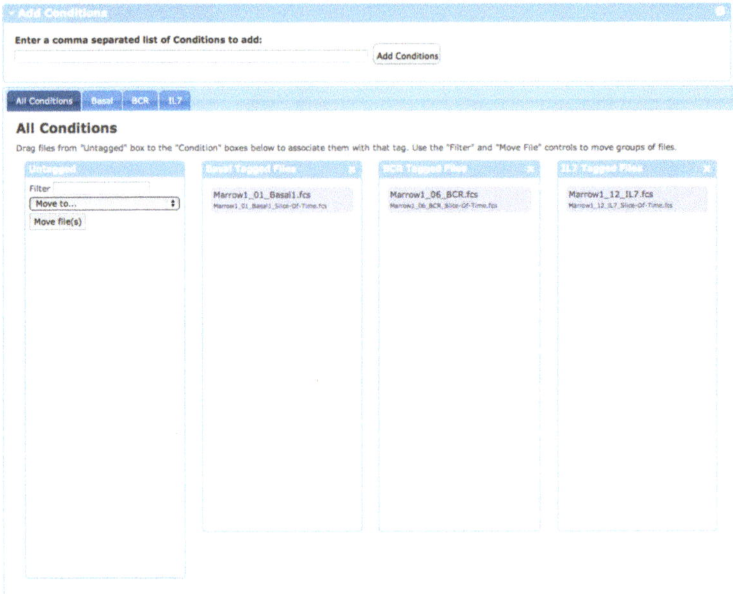


**Fig. 6** Automated annotation of conditions. After a researcher types in Basal, BCR, and Il7, FCS files are automatically categorized based on their original file annotations. This becomes especially useful in large experiments where many files are subjected to the same condition

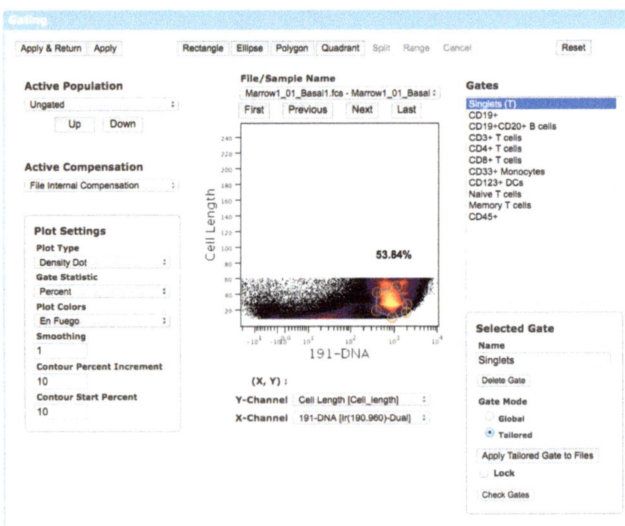


**Fig. 7** Example of gating a population in Cytobank. Users can view different plot types, change scales, create gates tailored for a specific file, view the population hierarchy, and even check that their gate is properly applied to all files by clicking on "Check Gate"

gate only one population prior to running SPADE, but multiple populations can be gated at this step. It is also important to check the plot scale settings on each channel to make sure that populations are clearly distinguishable. For more information about plots and scaling in Cytobank, visit the Cytobank Support and Blog sites (http://www.support.cytobank.org/ Cytobank Support).

For this dataset, we will gate first on DNA versus Cell Length to obtain the set of intact cells. Once each gate is made, select it in the gate list and click on the "Check Gate" button at the bottom right of the tool to ensure the gate is a good fit on all files. Once you are satisfied with the gate, click the "Apply and Return" button to go back to the Working Illustration. The gating tool is a scratch space where gates that you draw do not affect the current Working Illustration. "Apply and Return" changes gates for an entire Experiment.

## *5.4 Generating Illustrations*

### 5.4.1 Working with Plot Types in Figure Dimensions

On the left under Plot Controls, users have the options of many different plot types. There are also controls for plot size and for selection of X and Y dimensions in 2D plots. For this figure, we will set up a heatmap. Under "Plot Types" select "Heatmap." The "Update Illustration" button on the top left is now yellow, and a green notice has popped up. Select "Click here" to apply these illustration changes.

By harnessing the power of annotating these conditions, the data can be stratified by Conditions, Channels, and Populations very quickly. For example, by setting these as Columns, Rows, and first Table, respectively, and selecting a subset of the Channels (in this case, phosphorylated proteins), the overview of the data shown in Fig. 8 can be achieved. Switching the Columns and Rows rearranges the heatmap as shown in Fig. 9. Extending this principle to multiple doses, sample types, and other figure dimensions allows users to easily explore the underlying original data and communicate findings. Any changes to gates, scales, or other figure panels are reflected in the figures shared with all users.

### 5.4.2 Saving a Snapshot of Your Working Illustration

To share an illustration with a collaborator, a snapshot of the Working Illustration can be saved. Saved Illustrations cannot be edited but can be copied into a user's Working Illustration as the foundation for a new analysis.

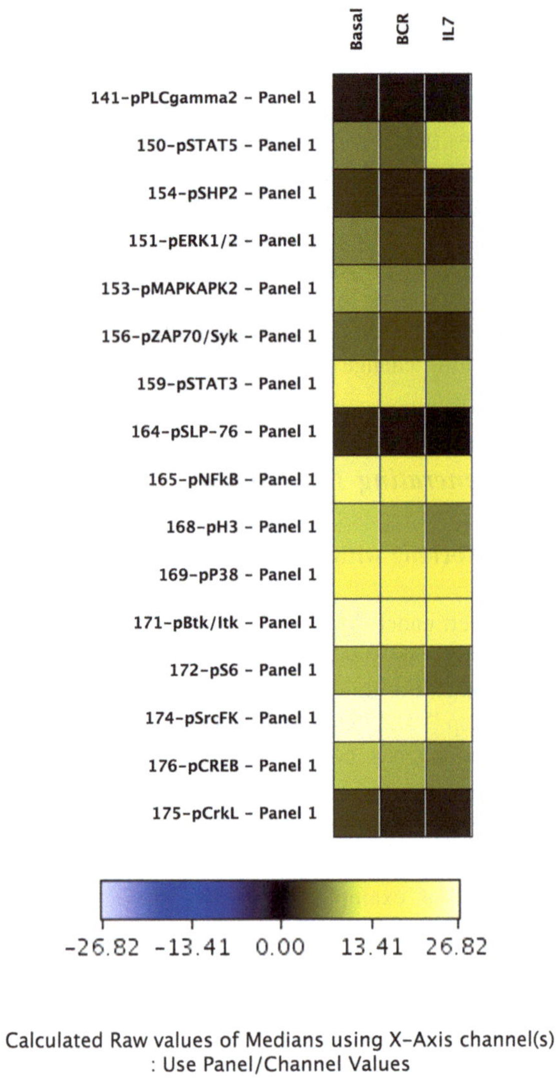


**Fig. 8** Example heatmap where the columns are the conditions and the rows are the channels

## 5.5 *SPADE Analysis*

### 5.5.1 Setting up SPADE

To perform a SPADE analysis, return to the Experiment Details page. Click "Create SPADE Analysis" under Actions on the left side of the page, at the bottom. Name this SPADE analysis with the date and the words "Spade Tutorial Science" and press OK. Setting up SPADE consists of five steps, where the items are displayed to the user in the SPADE interface. In this example, cell populations

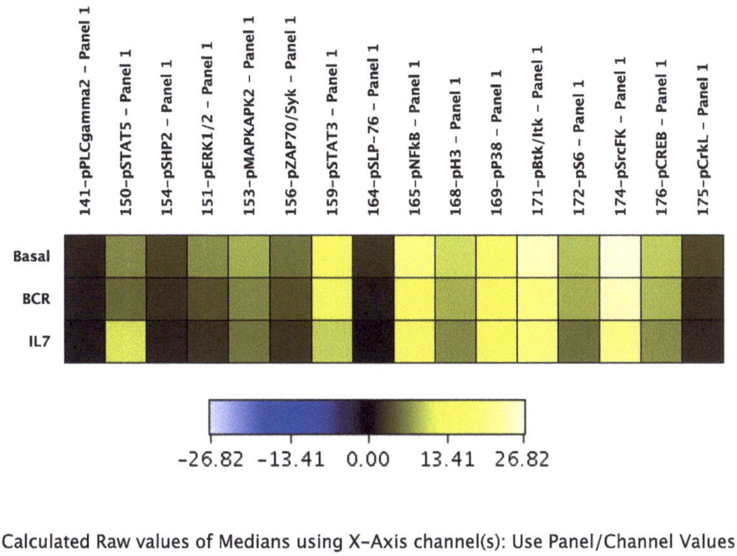


**Fig. 9** Example heatmap where the figure dimensions are switched relative to the heatmap shown in Fig. 8 such that the Conditions are the rows, and the Channels are the columns

have already been gated. If the SPADE analysis were being performed on fluorescence data, the applied compensation must be chosen.

*Target Number of Nodes*

- Select the approximate number of final nodes desired in the tree.
- Typically, higher numbers of nodes are used to find transition or rare cell populations.

*Downsampled Events Target*

- The default downsampling parameter is 10 %. This number can be decreased if you are looking for more rare cells.
- The target number can also be defined as an absolute number.

*Population*

- Select the population on which SPADE will be run.
- If the entire sample is to be analyzed, an intact cells gate is often sufficient.

*Clustering Channels*

- Clustering channels allows selection of the biological context for viewing of the data. In this example, we will be studying immunological cell types and thus will cluster on surface markers only.

**Fig. 10** Setting up SPADE groups. In this menu, different fold change groups are created and files are selected for the baseline calculation of each group

*Fold-Change Groups*

- To calculate fold change, the basal level for each group of samples is defined here.

For this example, the default Target Number of Nodes and default Downsampled Events Target will be selected. Under Population, choose the Singlets population for SPADE. For Clustering Channels filter and choose all CD markers except for the duplicate CD3 channels (110 through 114), leaving only the combined 110_114-CD3. Finally, click "Setup" under Fold-Change Groups. Click the checkbox for the Basal sample for Baseline (Fig. 10).

After a user clicks the start link to run the analysis, a progress bar appears. Because of the computational complexity of cluster analysis in high dimensions, a large amount of computing power is allocated for each SPADE run on the cloud rather than on the user's local machine. A user can close the browser window after beginning the analysis. The run will undergo a downsampling step, a clustering step, an upsampling step, and finally a median calculation prior to completion. When the SPADE job is complete the user will receive an email and can log on to access the calculated SPADE tree. An example of an unannotated tree is shown in Fig. 11. SPADE trees may vary in their layout, and are by default colored by cell count. The size of the node represents the number of cells.

### 5.5.2 Leveraging SPADE for Effective Communication of Analysis Through Bubbling and Annotation

Bubbling is defined as the manual grouping of SPADE nodes in the user interface. To help with bubbling, there is a 2D visualization tool on the left which will help you to view selected cells. When you click and drag to highlight nodes in the SPADE tree, these events will appear on the Events Plot. You can also click "Pop-Out New Plot" to visualize many 2D plots at the same time.

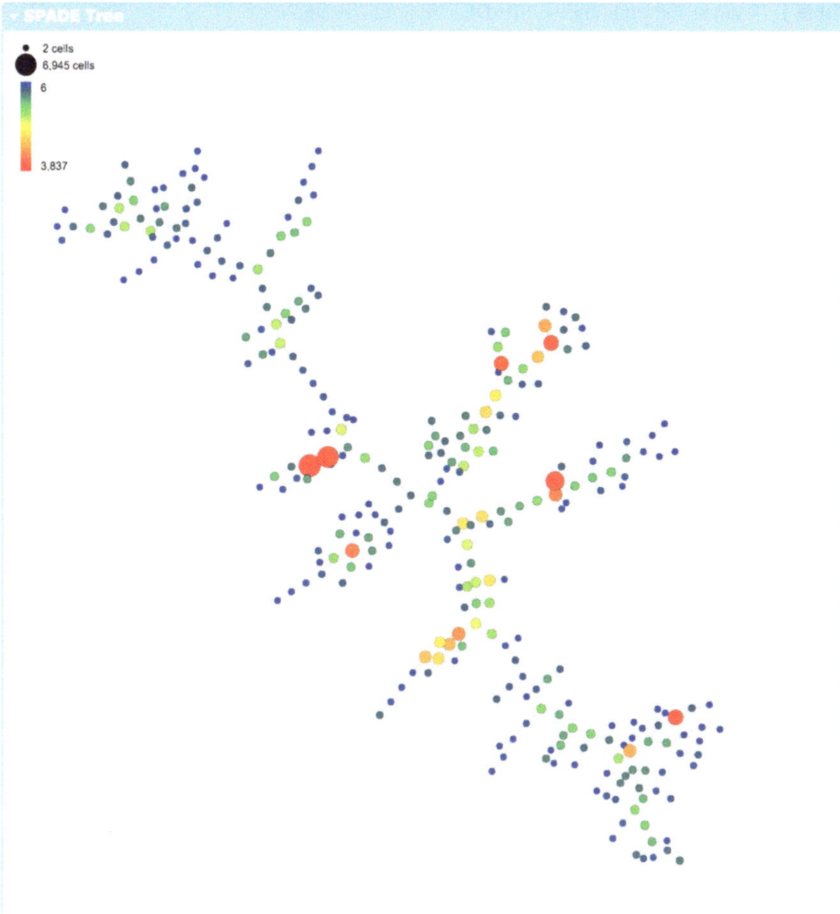


**Fig. 11** Example un-annotated SPADE tree. After a SPADE job is completed, a tree is displayed. Each node represents a cluster of cells. This tree is colored by the event count in each node, where red is high and blue is low. The size of each node is also proportional to the event count

In this exercise we will incrementally bubble groups of nodes. On the left side under SPADE Tree Rendering, select CD3. Drag to select the $CD3^+$ cells and then click "Add Bubble." Then color by CD4 and CD8 to distinguish those sets of nodes. Once you have completed this, then you can remove the $CD3^+$ bubble. A table below is given to help guide you through the bubbling process (Table 1).

You can highlight and drag a group of nodes to a different location or press "Z" to rotate them. Switching to fold change under "Metric" will also allow visualization of differences between samples. Right and left arrows are used to switch between files, and the up and down arrows switch between channels. The example

**Table 1** A guide for bubbling the SPADE tree in this example

| Markers | Population |
|---|---|
| $CD3^+$, $CD4^+$, $CD45RA^+$ | $CD4^+$ Naïve T cells |
| $CD3^+$, $CD4^+$, $CD45RA^-$ | $CD4^+$ Memory T cells |
| $CD3^+$, $CD8^+$, $CD45RA^+$ | $CD8^+$ Naïve T cells |
| $CD3^+$, $CD8^+$, $CD45RA^-$ | $CD8^+$ Memory T cells |
| $CD19^+$, $CD20^+$ | B cells |
| $CD33^+$ | Monocytes |
| $CD123^+$ | Dendritic cells |
| $CD34^+$ | Progenitor |

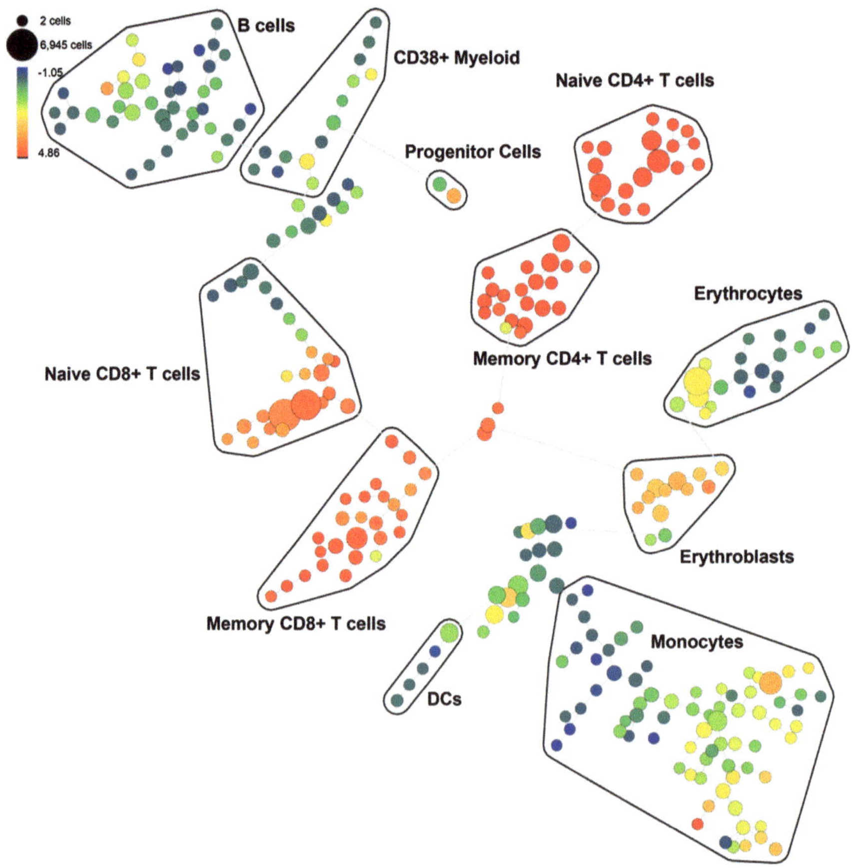


**Fig. 12** An annotated SPADE tree created using the bubbling guide. Note that this tree is colored by CD3, thus nodes containing T cells are colored red and orange whereas others are not

shown in Fig. 12 is a bubbled SPADE tree colored by CD3. Post annotation, bubbled events can be exported from SPADE as cell populations and can be viewed as heatmaps, histogram overlays, or 2D plots in Cytobank.

# 6 Conclusions and Future Directions

## *6.1 Transparency of Data Analysis is a Requirement for Systems Level Studies*

The trend in cytometry has been toward the development of technologies and methods that will enable researchers to measure increasingly more data in each experiment. These efforts will save time, money, and effort and maximize information yield from each valuable sample. Novel analysis tools have been and are being built to support exploration and analysis these high-dimensional datasets. The Institute of Medicine of the National Academies recently put forth recommendations surrounding the analysis of high-dimensional datasets. This was made necessary in part because the mismanagement of the analysis process can result in erroneous conclusions and has led to retraction of several published large-scale studies. The theme behind each of the recommendations is the need for transparency across all arms of the analysis process to facilitate verification and replication of published data. Researchers are strongly urged to make raw data, metadata, analysis strategies, and computational procedures "sustainably available" to "interested parties" involved in the process via independent repositories. Interested parties include "the biomedical and clinical research community, investigators, institutions—public and private, commercial and nonprofit—funders of omics research, and journals that publish the results of omics research and clinical trials" (Academies). Sustainability of preserving collected and annotated data is a key part of scientific research—currently, the availability of general research data declines rapidly as time progresses (Vines et al. 2014).

High-dimensional datasets are rich in information and complex in their analysis requirements. Consistency in analysis approach and transparency of the entire process will facilitate the generation of robust conclusions, as multiple interested parties will have the opportunity and ability to supervise the process and participate in moving forward from the point of publication. Cloud platforms such as Cytobank lend themselves to the centralization of data as datasets can be managed, shared, and analyzed from anywhere in the world and from any web-enabled device. Cytobank provides a number of sharing and collaboration-oriented functionalities, giving researchers the control to set data accessibility permissions. Cytobank Projects allow users to give collaborators access to datasets at selected levels of access. A number of Cytobank users have also opted to make their published datasets publicly available. Large datasets are likely to be analyzed and accessed by a number of individuals participating in the analysis process over time.

Providing a consistent framework for annotation also promotes consistency in data analysis strategies. Multi-center collaborative efforts have benefited from using a centralized analysis platform. Tools for importing and templating help obviate user errors in the annotation process. The annotation process in Cytobank involves layering information about samples ("annotations") on top of raw data. This includes information about stimulation conditions, time points, dosages, donors, populations, and other related information. These annotations are made available alongside the raw data, allowing others who can access the data to perform informed analyses. Collaborators can log in to Cytobank and interact with these analysis strategies independent of geographical location.

Educational resources linked to underlying data can also be built on top of these platforms further facilitating data sharing and analysis transparency. For example the Nolan Lab Signaling Based Cytometry Resource (http://cytobank.org/nolanlab/) contains Cytobank Reports presenting their published findings with links to the raw data and accompanying analyses. These Reports are publically available and allow interested parties to interact directly with the raw data in Cytobank.

## *6.2 Novel Platform Development Must Continue as Cytometry Advances*

Cytometry datasets continue to become more complex as the number of markers measured per cell and experimental variables increase due to advances in instrumentation and reagents. Currently CyTOF® technology enables collection of 40 parameters; it has the potential to measure 100. Mass cytometry has already been used to provide a systems level view of the immune system during healthy development in disease states, and after drug treatment (Bjornson et al. 2013). Crucial to this capability, however, is development of the appropriate informatics technologies, collaborative environments, and computational infrastructure to allow analysis of the large amount of data generated. Current methods in cytometry analysis rely on several manual steps and accurately automating these stages will be important for reproducible analyses.

The Cytobank platform addresses many of the challenges in managing and analyzing cytometry dataset by building on recent advances in cloud computing and virtualization and by enabling efficient use of large computing resources for centralized storage, analysis, and collaboration. Through Cytobank users can access, organize, analyze, and share their results with their group members. Data are stored centrally and securely and experiments are accessible using a web browser to those with appropriate privileges.

The Cytobank approach grew out of the need to collaborate and follow a line of investigation from a sample through the analysis to a clinically digestible summary. This approach is also central to storage and analysis of the large amounts of mass and flow cytometry data generated in labs and core facilities today. Cytobank includes analysis tools and visualizations developed to handle high-dimensional datasets and provides a scalable platform to perform these analyses for users anywhere in the world. Key layers of permissions and security facilitate collaborations and allow use in clinical settings. The accessibility of raw data and annotations coupled with the ease of generating analyses will encourage researchers to look at more parameters and therefore lead to discovery of new patterns in the data. The mission of Cytobank and similar efforts is to ensure access to transparent analyses and to facilitate dynamic visualizations of primary data in collaborations and publishing.

**Acknowledgments** The authors would like to thank J. Irish and P. Krutzik for continued discussions and G. Kraker for help with analyses. The Cytobank project has been funded in part by the NIH including NHLBI Contract No. HHSN268201300037C, NIGMS Grant No. GM096579 and NIAID Grant No. AI094929.

## References

Aghaeepour N, Nikolic R, Hoos HH, Brinkman RR (2011) Rapid cell population identification in flow cytometry data. Cytometry A 79:6–13

Aghaeepour N, Chattopadhyay PK, Ganesan A, O'Neill K, Zare H, Jalali A et al (2012) Early immunologic correlates of HIV protection can be identified from computational analysis of complex multivariate T-cell flow cytometry assays. Bioinformatics 28:1009–1016

America tCotUSo HITECH Act of 2009

Amir el AD, Davis KL, Tadmor MD, Simonds EF, Levine JH, Bendall SC et al (2013) viSNE enables visualization of high dimensional single-cell data and reveals phenotypic heterogeneity of leukemia. Nat Biotechnol 31:545–552

Bagwell CB (2005) Hyperlog-a flexible log-like transform for negative, zero, and positive valued data. Cytometry A 64:34–42

Bandura DR, Baranov VI, Ornatsky OI, Antonov A, Kinach R, Lou X et al (2009) Mass cytometry: technique for real time single cell multitarget immunoassay based on inductively coupled plasma time-of-flight mass spectrometry. Anal Chem 81:6813–6822

Bashashati A, Lo K, Gottardo R, Gascoyne RD, Weng A, Brinkman R (2009) A pipeline for automated analysis of flow cytometry data: preliminary results on lymphoma sub-type diagnosis. Conf Proc IEEE Eng Med Biol Soc 2009:4945–4948

Bendall SC, Simonds EF, Qiu P, Amir el AD, Krutzik PO, Finck R et al (2011) Single-cell mass cytometry of differential immune and drug responses across a human hematopoietic continuum. Science 332:687–696

Bjornson ZB, Nolan GP, Fantl WJ (2013) Single-cell mass cytometry for analysis of immune system functional states. Curr Opin Immunol 25:484–494

Bodenmiller B, Zunder ER, Finck R, Chen TJ, Savig ES, Bruggner RV et al (2012) Multiplexed mass cytometry profiling of cellular states perturbed by small-molecule regulators. Nat Biotechnol 30:858–867

CDC/National Center for Health Statistics, O. o. I. S. (2009) International classification of diseases, ninth revision (ICD-9). Retrieved 24 Dec 2013 from http://www.cdc.gov/nchs/icd/icd9.htm

Christine MM, Sharly JN, Gilbert SO (Eds) (2012) Evolution of translational omics: lessons learned and the path forward, The National Academies Press

Congress U (2009) American recovery and reinvestment act of 2009. Public Law (111–5):111

Finck R, Simonds EF, Jager A, Krishnaswamy S, Sachs K, Fantl W et al (2013) Normalization of mass cytometry data with bead standards. Cytometry A 83:483–494
Frelinger J, Kepler TB, Chan C (2008) Flow: Statistics, visualization and informatics for flow cytometry. Source Code Biol Med 3:10
Ge Y, Sealfon SC (2012) flowPeaks: a fast unsupervised clustering for flow cytometry data via K-means and density peak finding. Bioinformatics 28:2052–2058
Gentleman RC, Carey VJ, Bates DM, Bolstad B, Dettling M, Dudoit S et al (2004) Bioconductor: open software development for computational biology and bioinformatics. Genome Biol 5:R80
Gremse M, Chang A, Schomburg I, Grote A, Scheer M, Ebeling C et al (2011) The BRENDA tissue ontology (BTO): the first all-integrating ontology of all organisms for enzyme sources. Nucleic Acids Res 39:D507–D513
Hahne F, LeMeur N, Brinkman RR, Ellis B, Haaland P, Sarkar D et al (2009) flowCore: a Bioconductor package for high throughput flow cytometry. BMC Bioinformatics 10:106
http://www.cytobank.org/nolanlab/ Nolan Lab Signaling-Based (Fluorescence & Mass) Cytometry Resource
Immport. Retrieved 24 December 2014, from https://www.immport.niaid.nih.gov/
Irish JM, Hovland R, Krutzik PO, Perez OD, Bruserud O, Gjertsen BT et al (2004) Single cell profiling of potentiated phospho-protein networks in cancer cells. Cell 118:217–228
Kotecha N, Krutzik PO, Irish JM (2010) Web-based analysis and publication of flow cytometry experiments. Curr Protoc Cytom Chapter 10(Unit10):7
Lee JA, Spidlen J, Boyce K, Cai J, Crosbie N, Dalphin M et al (2008) MIFlowCyt: the minimum information about a flow cytometry experiment. Cytometry A 73:926–930
Linderman MD, Bjornson Z, Simonds EF, Qiu P, Bruggner RV, Sheode K et al (2012) CytoSPADE: high-performance analysis and visualization of high-dimensional cytometry data. Bioinformatics 28:2400–2401
Lo K, Brinkman RR, Gottardo R (2008) Automated gating of flow cytometry data via robust model-based clustering. Cytometry A 73:321–332
Newell EW, Sigal N, Bendall SC, Nolan GP, Davis MM (2012) Cytometry by time-of-flight shows combinatorial cytokine expression and virus-specific cell niches within a continuum of CD8+ T cell phenotypes. Immunity 36:142–152
NIST (2013) Federal information security management act (fisma) implementation project. Retrieved 24 Dec 2013, from http://www.csrc.nist.gov/groups/SMA/fisma/.
Parks DR, Roederer M, Moore WA (2006) A new “Logicle” display method avoids deceptive effects of logarithmic scaling for low signals and compensated data. Cytometry A 69:541–551
Pyne S, Hu X, Wang K, Rossin E, Lin TI, Maier LM et al (2009) Automated high-dimensional flow cytometric data analysis. Proc Natl Acad Sci U S A 106:8519–8524
Qian Y, Wei C, Eun-Hyung Lee F, Campbell J, Halliley J, Lee JA et al (2010) Elucidation of seventeen human peripheral blood B-cell subsets and quantification of the tetanus response using a density-based method for the automated identification of cell populations in multidimensional flow cytometry data. Cytometry B Clin Cytom 78(Suppl 1):S69–S82
Qiu P, Simonds EF, Bendall SC, Gibbs KD, Bruggner RV, Linderman MD et al (2011) Extracting a cellular hierarchy from high-dimensional cytometry data with SPADE. Nat Biotechnol 29:886–891
Roederer M, Treister A, Moore W, Herzenberg LA (2001) Probability binning comparison: A metric for quantitating univariate distribution differences. Cytometry 45:37–46
Shen-Orr SS, Tibshirani R, Khatri P, Bodian DL, Staedtler F, Perry NM et al (2010) Cell type-specific gene expression differences in complex tissues. Nat Methods 7:287–289
Smith B, Ashburner M, Rosse C, Bard J, Bug W, Ceusters W et al (2007) The OBO Foundry: coordinated evolution of ontologies to support biomedical data integration. Nat Biotechnol 25:1251–1255
Society, F. I. a. C. C. (2013) Flow standards. Retrieved 24 Dec 2013, from http://www.ficcs.org/data/flow-standards/.

Spidlen J, Moore W, Parks D, Goldberg M, Bray C, Bierre P et al (2010) Data file standard for flow cytometry, version FCS 3.1. Cytometry A 77:97–100
Spidlen J, Breuer K, Rosenberg C, Kotecha N, Brinkman RR (2012) FlowRepository: a resource of annotated flow cytometry datasets associated with peer-reviewed publications. Cytometry A 81:727–731
Spidlen J, Barsky A, Breuer K, Carr P, Nazaire MD, Hill BA et al (2013) GenePattern flow cytometry suite. Source Code Biol Med 8:14
Tung JW, Parks DR, Moore WA, Herzenberg LA, Herzenberg LA (2004) New approaches to fluorescence compensation and visualization of FACS data. Clin Immunol 110:277–283
UniProt C (2010) The Universal Protein Resource (UniProt) in 2010. Nucleic Acids Res 38:D142–D148
Vines TH, Albert AY, Andrew RL, Debarre F, Bock DG, Franklin MT et al (2014) The availability of research data declines rapidly with article age. Curr Biol 24:94–97
Walther G, Zimmerman N, Moore W, Parks D, Meehan S, Belitskaya I et al. (2009a) Automatic clustering of flow cytometry data with density-based merging. *Adv Bioinformatics* 686759
Walther G, Zimmerman N, Moore W, Parks D, Meehan S, Belitskaya I et al (2009b) Automatic clustering of flow cytometry data with density-based merging. Adv Bioinform 2009:7
Wheeler DL, Barrett T, Benson DA, Bryant SH, Canese K, Chetvernin V et al (2008) Database resources of the National Center for Biotechnology Information. Nucleic Acids Res 36:D13–D21
Wodak S, O'Neill K, Aghaeepour N, Špidlen J, Brinkman R (2013) Flow cytometry bioinformatics. PLoS Comput Biol 9:e1003365
Zare H, Shooshtari P, Gupta A, Brinkman RR (2010) Data reduction for spectral clustering to analyze high throughput flow cytometry data. BMC Bioinform 11:403

# Computational Analysis of High-Dimensional Flow Cytometric Data for Diagnosis and Discovery

**Nima Aghaeepour and Ryan Brinkman**

**Abstract** Recent technological advancements have enabled the flow cytometric measurement of tens of parameters on millions of cells. Conventional manual data analysis and bioinformatics tools cannot provide a complete analysis of these datasets due to this complexity. In this chapter we will provide an overview of a general data analysis pipeline both for automatic identification of cell populations of known importance (e.g., diagnosis by identification of predefined cell population) and for exploratory analysis of cohorts of flow cytometry assays (e.g., discovery of new correlates of a malignancy). We provide three real-world examples of how unsupervised discovery has been used in basic and clinical research. We also discuss challenges for evaluation of the algorithms developed for (1) identification of cell populations using clustering, (2) identification of specific cell populations, and (3) supervised analysis for discriminating between patient subgroups.

## Contents



N. Aghaeepour · R. Brinkman (✉)
Terry Fox Laboratory, BC Cancer Agency,
675 West 10th Avenue, Vancouver BC V5Z 1L3, Canada
e-mail: rbrinkman@bccrc.ca

Current Topics in Microbiology and Immunology (2014) 377: 159–175
DOI: 10.1007/82_2013_337


Published Online: 25 August 2013

# 1 Introduction

For more than 30 years, the fluorescence-based technique of flow cytometry (FCM) has been widely used by clinicians and researchers to distinguish different cell types from mixtures of cell populations based on the expression of cell type-specific markers. The great advantages of FCM include its relatively low cost, rapid turnaround time, and amenability to standardization (Keeney et al. 2004). The large amount of data generated by this technology poses unique informatics and statistical challenges (Boddy et al. 2001; Maino and Maecker 2004; Roederer and Hardy 2001; Bagwell 2004; Maecker et al. 2005; Herzenberg et al. 2002; Overton 1988; Roederer 2001; Suni et al. 2004). Critical problems in the use of FCM relate to (*1*) the design of the panels of biomarkers to be applied, (*2*) the evaluation of new versus 'classical' markers, (*3*) the analysis of the data obtained from the FCM measurements, and (*4*) the interpretation of results (Kalina et al. 2012). To address these issues, bioinformatics and computational biology groups have been developing computational solutions to address each step in the data analysis pipeline in order to transform the huge amount of information generated by high-dimensional FCM studies into summaries that are brief enough for creative studies by a human researcher (Fig. 1).

# 2 Platforms and Software

Using FCM as a high-throughput technology requires suitable software infrastructure to facilitate automated data handling. There are currently no commercial FCM processing packages that out-of-the-box fully accept multifactorial formats and allow rapid processing of large number of samples, keeping the metadata and relationships between samples intact (Robinson et al. 2012). As a result, open-source software development has been critical to the advancement of computational analysis of FCM data. The most well-developed framework is the free, open-source computational platform within the R/BioConductor statistical programming environment (Gentleman et al. 2004) that enables bioinformaticians, computer scientists, and statisticians to work collaboratively with biologists and clinicians to efficiently analyze FCM data, a process deemed crucial by many for

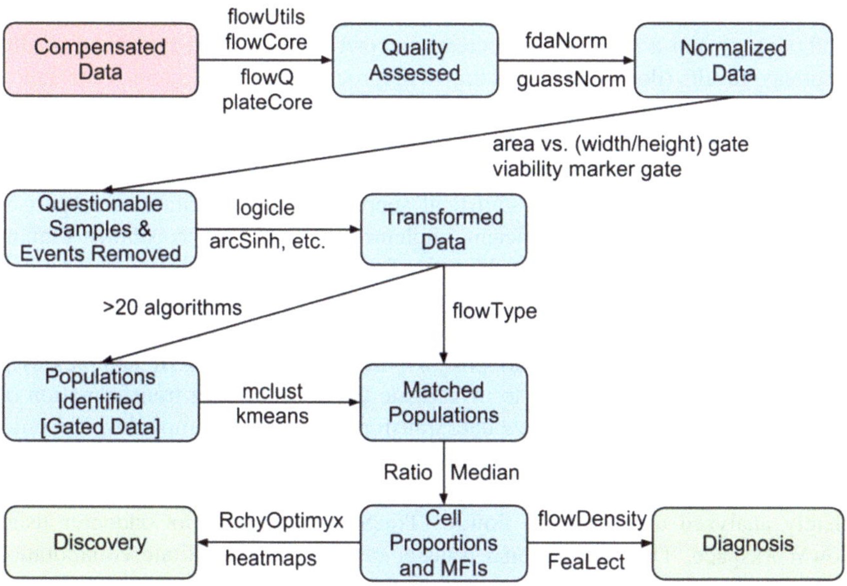


**Fig. 1** BioConductor-based, modular high-throughput flow cytometry data analysis workflow

the further development of FCM technology (Lizard 2007). The platform is based on a modular architecture that enables developers to extend and use the underlying infrastructure and to combine tools in complex data analyses (Le Meur 2013). This has been a success given more than 20 individuals from 11 groups being the primary maintainers of high-throughput FCM data analysis packages available through BioConconductor (Bioconductor 2013) and other sites.

The framework is based on the flowCore package that provides data structures and standardized tools designed for exploration and data analysis of FCM data analysis (Hahne et al. 2009b). All related FCM packages, including those for data analysis, visualization, and quality control, use the flowCore infrastructure to access data. Objects representing individual or collections of FCM samples are memory-resident data structures; entire datasets are maintained in RAM. The basic memory model for FCM data cannot handle the massive datasets (millions of cells, hundreds or thousands of samples) generated by high-throughput instruments and clinical trials. This limitation poses a significant difficulty for scientists and bioinformaticians who use BioConductor tools for massive FCM/MCM data analysis, even on high-end workstations. In order to analyze such large-scale clinical datasets, ncdfFlowSet inherits most of the data structures from the existing flowSet object model. It stores the large volume of event-level data on disk and keeps only the file handle and metadata in memory. Thus, memory consumption is significantly reduced. Data are stored on disk in NetCDF format. This format is self-describing, machine-independent, and specifically optimized for storing and accessing array-oriented scientific data, allowing users to analyze gigabytes of FCM data on a regular workstation with a modest amount of RAM.

The BioConductor framework also includes advanced quality assurance methodology and a framework to create interactive web-based reports of quality assurance results (flowQ), mechanisms to preprocess data files to remove technical between-sample variation in FCM data through normalization (warpSet and gaussNorm (Hahne et al. 2009a)), and methods to estimate optimal data transformation for gating and visualization (flowTrans (Finak et al. 2010)). The flowStats package includes several statistical approaches for quantitative comparison of FCM data including an efficient implementation of the probability binning approach (Roederer et al. 2001). Methods for visualizing multiple FCM datasets have also been incorporated into the existing flowViz (Sarkar et al. 2008) package. In addition, several methods in flowQ can also be used to visualize and compare gated data (in one and two dimensions) across multiple FCM samples. The flowTrans feature allows users to investigate the effect of data transformation on automated gating and to optimize data transformation at the sample level in order to facilitate analysis, visualization, and quantitative comparison of FCM data from different samples or patients. Computational biologists are able to share completely analyzed data between flowJo (TreeStar Inc.) and BioConductor using flowWorkspace. This bidirectional transfer of data helps facilitate collaboration and communication between the computational research community and FCM end-users. Importantly, simplified access to manually gated data allows computational researchers to compare "gold standard" manual analysis against novel automated gating approaches and other sophisticated algorithms, reducing development time of new computational tools.

The GenePattern Flow Cytometry Suite (GenePattern.org; (Reich et al. 2006)) brings many of these advanced FCM data analysis tools from BioConductor to experimentalists without programming skills. It contains 34 open-source GenePattern FCM modules that can help with quality assessment, normalization, outliers removal, gating/clustering, cluster labeling, feature extraction, and other tasks. Using the GenePattern web-based interface, members of the genomics community can also connect these modules to build analytical pipelines. The openCyto data analysis framework also provides a simple-to-use interface to analysis algorithms available within R/BioConductor.

# 3 Cell Population Identification

## *3.1 Algorithms for Identification of Cell Populations*

For most high-dimensional single-cell assays, one of the first analysis tasks is the identification of phenotypically or functionally homogeneous "cell populations". For small fluorescence-based FCM datasets this can be achieved by simply using a series of scatter plots. An expert analyst first determines the order of the plots and then sequentially identifies the regions of interest by drawing polygons around them. Many problems have been noted with this approach to FCM data analysis, including its subjective, time-consuming nature, and the difficulty in effectively analyzing high-dimensional data (the number of required scatter plots that need to be investigated grows exponentially as the number of parameters increases). Manual analysis has been repeatedly identified as the largest source of variation in multicenter studies (Maecker et al. 2005; Lugli et al. 2010; Levin et al. 2013).

Beginning in 2008, several sophisticated clustering algorithms were developed to automatically identify cell populations in FCM data (e.g., (Lo et al. 2008; Aghaeepour et al. 2011; Ge and Sealfon 2012)). In theory, these algorithms are superior to manual analysis as they are more reproducible and can simultaneously consider all markers. Many different approaches have been used to identify populations, often relying on abstract definitions of clusters from set theory, information theory, or other branches of mathematics. In the absence of theoretical proof that FCM data follows specific mathematical patterns or distributions, choosing a clustering algorithm has remained a subjective task and has motivated empirical evaluation of these algorithms. Due to lack of high-quality public datasets, algorithms have often been evaluated on relatively small datasets and cross-comparison of performance has usually not been possible.

## *3.2 FlowCAP: Critical Assessment of Cell Population Identification Algorithms*

Guidance on the relative merits, appropriate use, and application of automated analysis methods is scarce. In response to this need, the FlowCAP effort seeks to advance the development of computational methods for the identification of cell populations of interest in FCM data by providing a way to objectively cross-compare them. The FlowCAP-I project is a collaborative effort for evaluation of cell population identification algorithms against the current best practice—manual analysis (Aghaeepour et al. 2013). The first FlowCAP set of challenges included five datasets analyzing 12–13 samples, 3–8 markers, and 5,000–1,00,000 cells. Larger datasets were not considered as their complete manual analysis was not feasible. Each dataset was manually gated by expert analysts. Cells that were not included in any of the manual gates were excluded from the evaluation but were still provided to the algorithms. Similarly, a wide range of cell population algorithms were used to assign each cell to a cell population automatically. These two sets of labels were then compared using their F-measure scores. The F-measure values revealed variation in the performance of the algorithms across different datasets (see Table 2 of (Aghaeepour et al. 2013)). The cell populations found by some algorithms were more similar to the manual gates than were those found by others. Interestingly, the consensus of all cell populations (produced using an ensemble clustering algorithm) had the highest F-measure across all algorithms. Sequential removal of algorithms from the consensus revealed that only a small subset of all algorithms is needed in the consensus to maintain the high F-measure value (Supplemental Fig. 6 of (Aghaeepour et al. 2013)). Finally, to investigate the sensitivity of the F-measure values two of the datasets were analyzed by eight independent human analysts. The results remained consistent with the evaluations using the original manual gates. Also, the consensus of several algorithms (as noted above) was shown to be most similar to a computationally calculated consensus of the eight human analysts.

## *3.3 Use of FCM in the Real-World*

High-dimensional FCM is often used for exploratory analysis of heterogeneous samples to identify cell populations correlated with an external factor (e.g., a clinical outcome). Most clustering tools for cell population identification perform an "exhaustive" clustering (i.e., attempt to identify all cell populations) and, therefore, in theory are suitable for these exploratory studies. In practice, this requires labeling these cell populations across multiple samples using a cluster matching procedure. These matching algorithms have proved to be both computationally challenging (similar to the "generalized assignment problem," an NP hard combinatorial problem) and subjective (e.g., some samples might not have all cell populations or some cell populations might not match very accurately) (Pyne et al. 2009). A partial solution to these challenges is to pool all of the samples (or a representative of all their clusters) into a single high-dimensional space for a secondary "meta-clustering" step. These meta-clusters can then be used as cross-sample "features" for correlative studies. However, this approach is sensitive to technical or biological variations that affect the median expression values of the clusters. In addition, in these exploratory settings the computational pipelines often identify complex cell populations involving many (if not all) markers, making biological interpretation and validation of the results challenging. Here we provide an overview of our flowType/RchyOptimyx pipeline followed by several examples of a computational pipeline that addresses these issues (Aghaeepour et al. 2012a, b).

## *3.4 flowType and RchyOptimyx*

The flowType algorithm starts by assigning all cells into a negative or a positive partition based on each marker. This simple strategy enables the use of any of the clustering algorithms mentioned earlier or, in case of rare and "shoulder" populations, manual gates provided by an expert analyst. These one-dimensional gates are then combined to produce high-dimensional cell populations. To enable our statistical model to objectively identify and exclude redundant markers, flowType also measures combinations of one-dimensional partitions that do not include all markers. Therefore, each marker can be positive, negative, or "neutral."

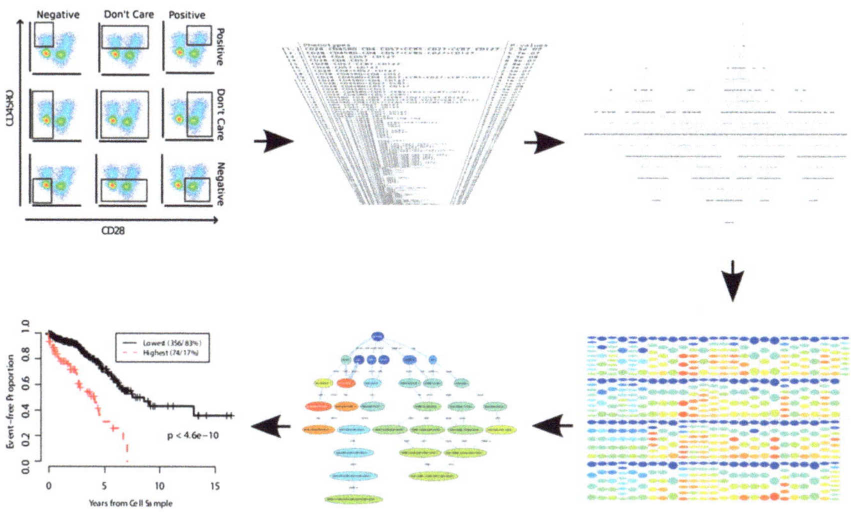


**Fig. 2** An overview of the flowType/RchyOptimyx pipeline. One-dimensional gates are combined to produce high-dimensional cell populations. A statistical test is used to generate a "hit list" of statistically significant phenotypes. For each population in the hit list a complete hierarchy of all parent populations is generated, then optimized, and finally merged into a single hierarchy representing the entire hit list. The hierarchy can be used to identify cell populations correlated with the external outcome

Therefore, for $M$ markers a total of $3^M$ phenotypes are measured. Figure 2 provides an overview of the flowType/RchyOptimyx pipeline.

The cell populations identified can then be analyzed using an appropriate statistical test (depending on the outcome variable being measured), subject to proper multiple hypothesis testing correction and sensitivity analysis, to produce a hit list of cell populations with a statistically significant correlation with the external outcome. Many of these cell populations will be highly correlated with each other and, as a result of including a "neutral" state, will overlap and have several common parent populations (e.g., $CD3^+CD4^+CD57^+$ and $CD3^+CD4^+CD8^-$ have $CD3^+$, $CD3^+CD4^+$, and $CD4^+$ cells as their parent populations). An example of a correlation heatmap is shown in Fig. 2 of (Aghaeepour et al. 2012b).

To identify the best representation of these hit lists, we developed RchOptimyx an algorithm for efficiently evaluating the "impact" of each marker (or combination of markers) on the statistical significance of each phenotype. RchOptimyx organizes all phenotypes in the hit list into a single hierarchy based on the most significant parent populations (see Fig. 5 of (Aghaeepour et al. 2012b) for a complete hierarchy for a relatively large dataset).

# 4 Analysis of FCM Data: Real-World Examples

## *4.1 Example 1: Mapping Signaling Pathways to Surface Markers*

As our first example, we provide an analysis of a publicly available data from a mass cytometry-based analysis of signaling pathways in healthy human bone marrow (data available through (Bendall et al. 2011)). The goal of this study was to identify a phenotype based on surface markers only that best identified the response to external stimulation as measured by the respective pathways. This was a proof of concept for fluorescence-activated cell sorting (FACS) isolation of live cells based on signaling pathways without permeabilization for intracellular staining. The dataset consisted of analyses of samples stimulated by IL3, BCR, and LPS as well as a basal control. STAT5, pBLNK, and p38 were used for identification of the responding cells. We measured the overlap of each cell population (CP) with the responding cells after background subtraction. For example, for IL3:

$$Overlap^{IL3} = \left(\frac{\#IL3^{+} cells\ in\ CP}{\#cells\ in\ CP}\right)_{Stim} - \left(\frac{\#IL3^{+} cells\ in\ CP}{\#cells\ in\ CP}\right)_{Unstim}$$

The immunophenotypes with a high overlap with the response population were selected for RchyOptimyx analysis (see Supplemental Tables S1–S3 and Supplemental Figure S6 of (Aghaeepour et al. 2012b) for details). The final hierarchies are shown in Fig. 3. These results are generally in agreement with the previous analysis (Bendall et al. 2011) of the same dataset with a few exceptions. For example, $CD33^{+}$ cells were not previously shown to be affected by IL7; however, the RchyOptimyx analysis revealed the importance of CD33 within the $CD3^{+}CD4^{+}$ compartment (Figs. 3 and 4). This demonstrates the ability of RchyOptimyx to perform a deep analysis to detect rare cell populations that were not identified using dimension reduction and other visualization methods.

## *4.2 Example 2: Characterization of Phenotypic Changes Induced by Genetic Alterations in Vitro*

This example demonstrates a real-world use of the flowType/RchyOptimyx pipeline for hypothesis generation in a basic research setting. Genetic alterations in the gene encoding Ikaros predicts poor outcome in leukemia. However, the role of Ikaros in normal and leukemic cells is unknown. We used flowType and RchyOptimyx to analyze the effect of disrupted Ikaros activity in cord blood cells. To produce this dataset, the green fluorescence protein (GFP) and a mutant form of Ikaros (IK6) were co-expressed and cultured with control cells marked with the

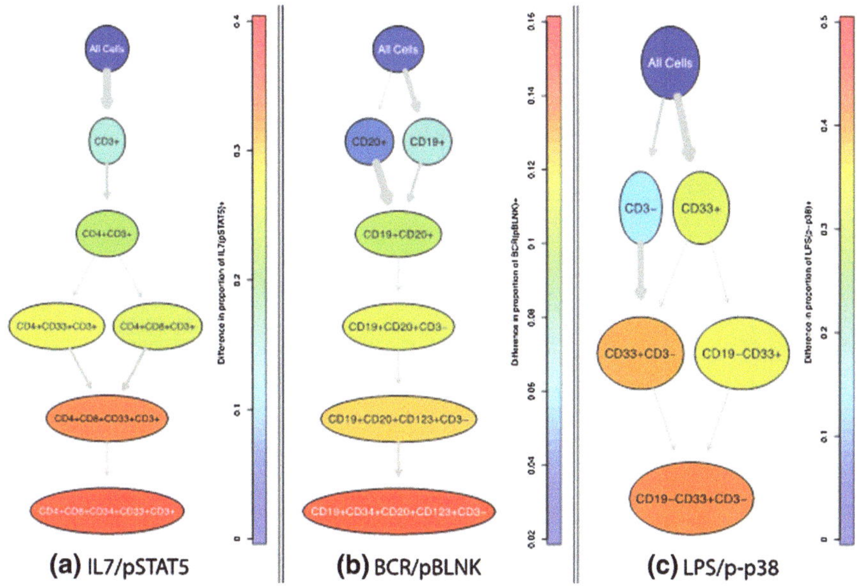


**Fig. 3** RchyOptimyx plots for identification of cells responding to three stimulations using surface markers only. Reproduced from (Aghaeepour et al. 2012b)

yellow fluorescence protein (YFP). After 48 h in culture, FCM was performed to measure levels of CD34, CD38, CD123, CD45RA, CD135, CD90, and CD131. The cell populations measured by flowType were scored based on their enrichment of GFP cells. Large cell populations (more than ~1,250 cells) with more than ~ 90 % GFP were selected for RchyOptimyx analysis (Fig. 5a, b). Similar results were produced using much less strict thresholds (Fig. 5c, d). However, these hierarchies included many other closely related populations and made visualization difficult. The identified cell population (Fig. 5d) perfectly matched the known signature of primitive stem cells (CD34 + CD38-CD45RA-CD90 +). These results justified the design of further in vivo experiments for validation of the effect of Ikaros on primitive stem cells.

### *4.3 Example 3: Improving Flow Cytometric Assessment of Germinal Center B Cell Lymphoma*

In this final example, we used the flowType/RchyOptimyx pipeline in a clinical setting to design a strategy for discriminating between germinal center B cell lymphoma (GC-L) and lymphoid hyperplasia (GC-H). The dataset consisted of analyses of samples from 52 and 48 GC-L and GC-H patients, respectively.

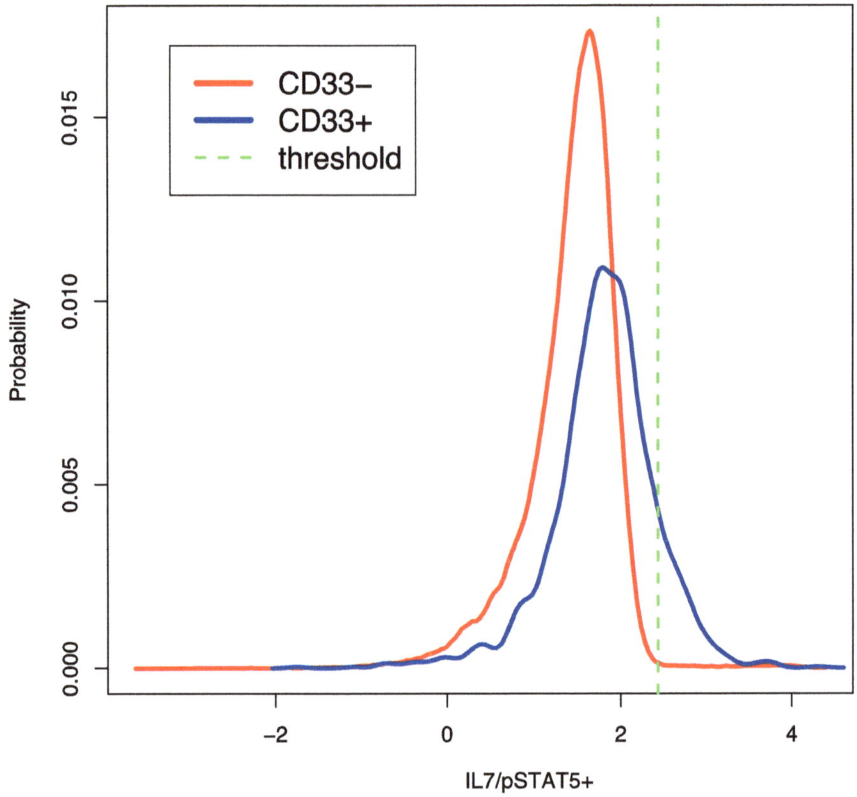


**Fig. 4** Kernel density estimation of IL7/pSTAT5$^+$ cells for the CD33$^-$ and CD33$^+$ cells in the CD3$^+$CD4$^+$ compartment

The panel included the following eight markers: KAPPA, LAMBDA, CD5, CD19, CD10, CD38, CD20, and CD45. To facilitate biological interpretation of the results, each marker was manually partitioned (Fig. 6a) as opposed to the previous examples in which a clustering algorithm was used. ROC analysis was performed on the $3^8$ (6,561 total) measured immunophenotypes and those with an AUC of higher and 0.9 were selected for RchyOptimyx analysis (Fig. 6b). RchyOptimyx revealed several cell populations that discriminated between the two diseases (Fig. 6c). As an example, the box plots for the CD5$^-$CD19$^+$CD10$^+$CD38$^-$ cell population are shown in Fig. 6d. These findings were successfully validated through independent manual analysis of the same dataset.

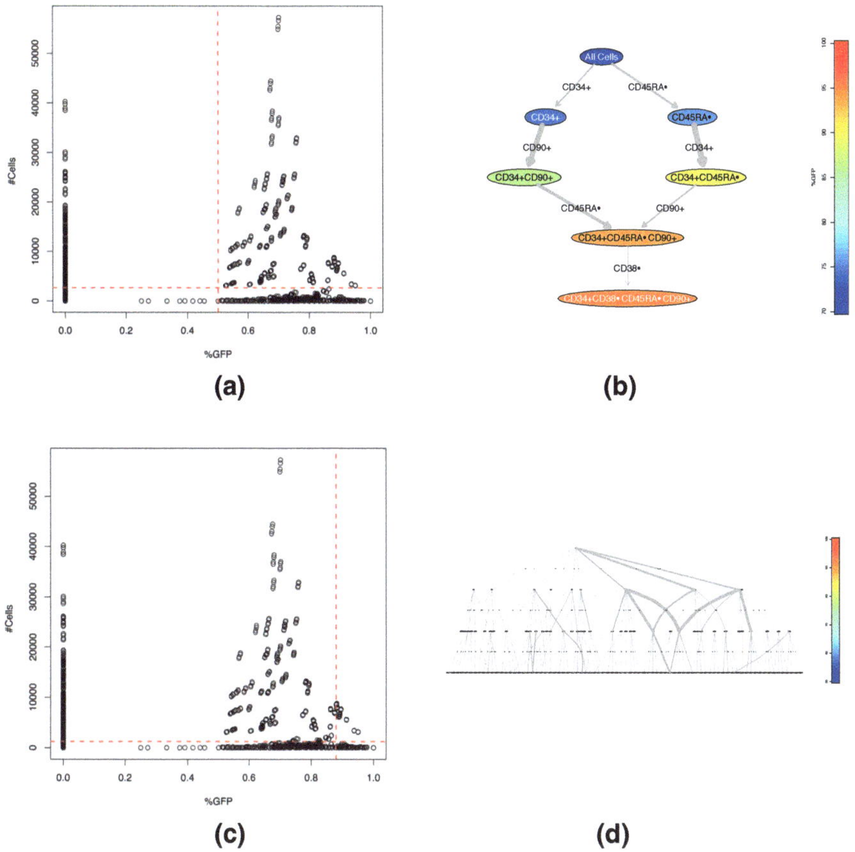


**Fig. 5** **a** The proportion of GFP cells versus the number of cells in each cell population. Only the cell populations with the highest GFP enrichment are selected for further analaysis. **b** RchyOptimyx analysis of the cells in the top right quadrant to reveal the most important parent populations **c** different thresholds to include larger populations with less enrichment for GFP + cells **d** RchyOptimyx analysis for the cell populations selected in panel (c)

# 5 Issues with Untrained Algorithms

## *5.1 Semi-supervised and Supervised Population Identification*

An untrained algorithm cannot substitute for the expertise of a clinician or a researcher when specific cell populations of interest are rare. However, many applications (e.g., clinical trials, high-throughput screenings) are burdened by the manual extraction of information (i.e., repetitive gating of identical populations) in a manner that is reproducible and robust across replicates. Despite the high overall

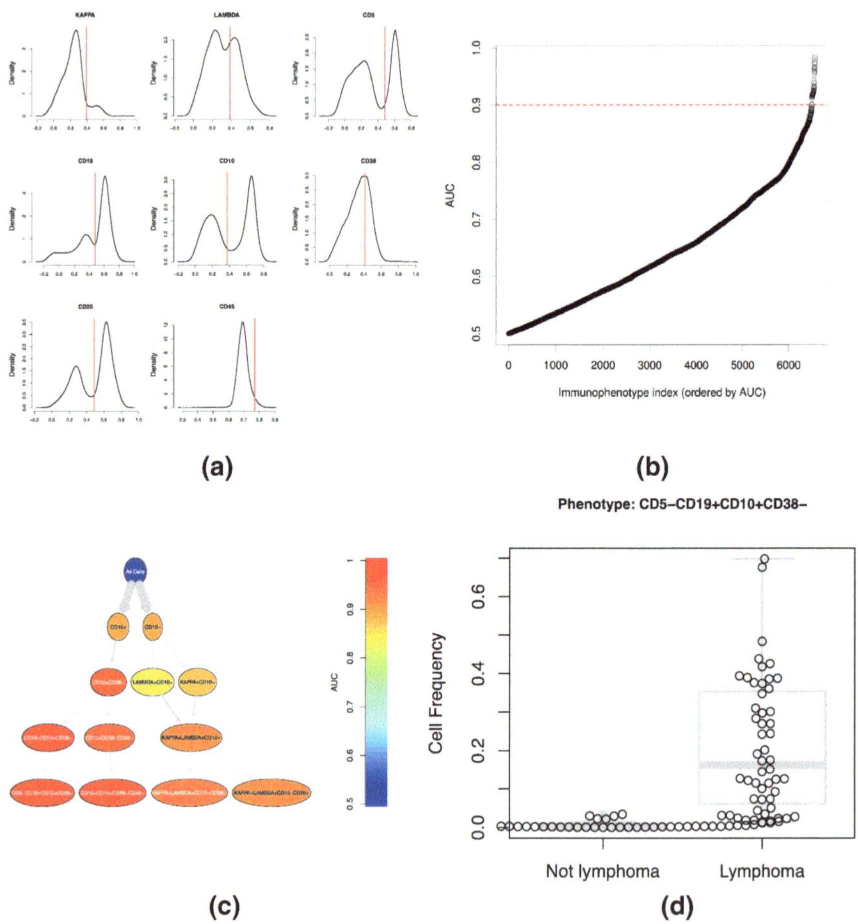


**Fig. 6** **a** Manual gates used for partitioning each marker. **b** The distribution of the AUCs of all immunophenotypes. **c** RchyOptimyx tree for the immunophenotypes above the red line in the previous panel. **d** Example boxplots of the proportions of cells for one of the selected immunophenotypes

ability of unsupervised automated clustering algorithms to match manual gating, they often fail to replicate a human expert's results for rare cell populations. They often return a number of clusters that cannot be interpreted readily without more sophisticated cluster matching methods. In particular, if the goal is to identify predefined target cell populations, these algorithms fail to deliver straightforward results that can be readily examined by human experts in two dimensions and are mainly suitable for discovery purposes where no particular cell population of interest exists a priori. The FlowCAP-I project provided preliminary data suggesting that when the specific target cell populations need to measured, algorithms can be guided by example manual gates for more accurate results.

## *5.2 Multivariate Classification Using Several Cell Populations*

As described in the previous section, computational pipelines can extract millions of phenotypical and functional cell types from a high-dimensional FCM dataset. These can be combined using multivariate classifiers to automatically diagnose a disease (or the subtype of the disease) very accurately (e.g., (Zare et al. 2012)). However, this often results in a high false discovery rate unless the cohort is large and diverse enough to cover the real heterogeneity of the disease of interest, and the findings must be validated in an independent dataset (Jelizarow et al. 2010).

The FlowCAP-II set of challenges, in collaboration with the DREAM initiative, evaluated a large number of multivariate classifiers on two relevant, large datasets (Aghaeepour et al. 2013). For both of these datasets, half of the cohort was provided to the participants for training purposes and the other half was used for an independent validation. The first dataset consisted of eight panels of different markers for analysis of 316 acute myeloid leukemia (AML) and 43 healthy subjects (more than 2,800 FCS files). As expected, the differences between AML and healthy subjects were quite large and most of the algorithms were able to correctly diagnose more than 90 % of the AML patients. The second dataset was provided by the HIV Vaccine Trial Network (HVTN) and consisted of post-HIV vaccination T cells stimulated by two different antigens. The majority of the evaluated multivariate classifiers were able to accurately distinguish between the two antigen stimulations. The cell populations selected by most algorithms included $CD4^{+}IL2^{+}$ cells, a cell population also identified in previous manual analysis of the same dataset (Env-stimulated samples have more of these cells). However, the accuracy of the predictions based on manual analysis was lower than that of the automated classifiers. Upon further inspection of the dataset, a possible run-specific variation was observed in the dataset. Excluding the samples that were affected by this increased the accuracy of the manual analysis to match that of the multivariate classifiers. This demonstrated the superiority of computational tools for diagnosis using several cell populations in presence of technical variation.

# 6 The State of the-Art and Unmet Needs

Through R/BioConductor, users can freely access to all the tools needed to implement an algorithmic pipeline for diagnosis or discovery. There are still several remaining challenges, such as direct quantitative comparison of multidimensional gated data, where gates are defined using three or more dimensions. This is due to the fact that gates extracted from automated algorithms are unlabeled and need to be matched across samples before populations statistics extracted from these gates can be compared (Finak et al. 2010). Unfortunately, this is not straightforward with automated gating as (1) gates are defined in high dimensions and (2) automated algorithms can return different number of populations per sample. Some attempts (Pyne et al. 2009) have been made to match and label populations automatically using their mean fluorescence intensity (MFI), but their application has been limited due to technical variation that can result in significant MFI changes across samples.

Another important barrier to the adoption of automated tools is their lack of biological understanding of the data they analyze. Clustering methods currently have no way to automatically apply meaningful labels (e.g., T cells) to cell populations they identify. The hematopoietic branch of the cell ontology (Diehl et al. 2011) has the potential to provide the foundation for providing such semantic tags. The ontology preferentially uses cell surface protein expression as the main characteristics for defining cell types in order to be able to link the terminology with experimental data and is especially designed to work with FCM. Such mappings would facilitate statistical analysis of cell populations as potential biomarkers in large cohort studies across institutions. They would also connect the results of FCM analysis methods with the wealth of knowledge about cellular processes and functions available in the scientific literature (i.e., through Gene Ontology (GO) (Gene Ontology Consortium 2004)) and emerging database resources like the Immunology Database and Analysis Portal (ImmPort) for NIAID-supported studies, CytoBank, FlowRepository (for datasets associated with peer-reviewed publications (Spidlen et al. 2012a, b)), and the Immune Tolerance Network TrialShare clinical trials research portal.

The tenant of garbage-in–garbage-out is very relevant to highly automated data processing of large datasets, where it is not possible to manually review each file for individual variation. Although quality checking approaches are available, steps to reduce sample variation greatly enhance the applicability of automated analysis. Groups such as EuroFlow (Kalina et al. 2012) and the Human Immunophentophing Consortium (Maecker et al. 2012) are moving toward making the data generated by studies standardized though the use of set reagent panels for use across multiple centers and are investigating the use of lyophilized reagents to further remove sample-to-sample variation.

As FCM is a ubiquitous assay in industry and biomedical research, the nascent field of automated analysis can have a large impact by reducing the need for manual analysis and increasing reproducibility. The promise of the combination of large collections of standardized datasets becoming routinely available, along with the tools to analyze this data in aggregate across centers, has the potential to revolutionize FCM and all the communities the technology serves.

## References

Aghaeepour N, Chattopadhyay PK, Ganesan A, O'Neill K, Zare H, Jalali A, Hoos HH, Roederer M, Brinkman RR (2012a) Early immunologic correlates of HIV protection can be identified from computational analysis of complex multivariate T-cell flow cytometry assays. Bioinformatics 28:1009–1016. doi: 10.1093/bioinformatics/bts082

Aghaeepour N, Finak G, Consortium The FlowCAP, Dougall D, Khodabakhshi AH, Mah P, Obermoser G, Spidlen J, Taylor I, Wuensch SA, Bramson J, Eaves C, Weng AP, Iii ES, Ho K, Kollmann T, Rogers W, De Rosa S, Dalal B, Azad A, Pothen A, Brandes A, Bretschneider H, Bruggner R, Finck R, Jia R, Zimmerman N, Linderman M, Dill D, Nolan G, Chan C, Khettabi FE, O'Neill K, Chikina M, Ge Y, Sealfon S, Sugar I, Gupta A, Shooshtari P, Zare H, De Jager PL, Jiang M, Keilwagen J, Maisog JM, Luta G, Barbo AA, Majek P, Vilcek J, Manninen T, Huttunen H, Ruusuvuori P, Nykter M, McLachlan GJ, Wang K, Naim I, Sharma G, Nikolic R, Pyne S, Qian Y, Qiu P, Quinn J, Roth A, The DREAM, Consortium Meyer P, Stolovitzky G, Saez-Rodriguez J, Norel R, Bhattacharjee M, Biehl M, Bucher P, Bunte K, Di Camillo B, Sambo F, Sanavia T, Trifoglio E, Toffolo G, Dimitrieva S, Dreos R, Ambrosini G, Grau J, Grosse I, Posch S, Guex N, Keilwagen J, Kursa M, Rudnicki W, Liu B, Maienschein-Cline M, Manninen T, Huttunen H, Ruusuvuori P, Nykter M, Schneider P, Seifert M, Strickert M, Vilar JM, Hoos H, Mosmann TR, Brinkman R, Gottardo R, Scheuermann RH (2013) Critical assessment of automated flow cytometry data analysis techniques. Nat Methods. doi:10.1038/nmeth.2365

Aghaeepour N, Jalali A, O'Neill K, Chattopadhyay PK, Roederer M, Hoos HH, Brinkman RR (2012b) RchyOptimyx: cellular hierarchy optimization for flow cytometry. Cytometry A 81:1022–1030. doi:10.1002/cyto.a.22209

Aghaeepour N, Nikolic R, Hoos HH, Brinkman RR (2011) Rapid cell population identification in flow cytometry data. Cytometry A 79:6–13. doi:10.1002/cyto.a.21007

Bagwell CB (2004) DNA histogram analysis for node-negative breast cancer. Cytometry 58A:76–78

Bendall SC, Simonds EF, Qiu P, Amir eD, Krutzik PO, Finck R, Bruggner RV, Melamed R, Trejo A, Ornatsky OI, Balderas RS, Plevritis SK, Sachs K, Pe'er D, Tanner SD, Nolan GP (2011) Single-cell mass cytometry of differential immune and drug responses across a human hematopoietic continuum. Science 332:687–696. doi:10.1126/science.1198704

Bioconductor (2013) Bioconductor—Flow Cytometry. http://www.bioconductor.org/packages/2.12/BiocViews.html#–FlowCytometry. Accessed June

Boddy L, Wilkins MF, Morris CW (2001) Pattern recognition in flow cytometry. Cytometry 44:195–209

Diehl AD, Augustine AD, Blake JA, Cowell LG, Gold ES, Gondre-Lewis TA, Masci AM, Meehan TF, Morel PA, Nijnik A, Peters B, Pulendran B, Scheuermann RH, Yao QA, Zand MS, Mungall CJ (2011) Hematopoietic cell types: prototype for a revised cell ontology. J Biomed Inform 44:75–79. doi:10.1016/j.jbi.2010.01.006

Finak G, Perez JM, Weng A, Gottardo R (2010) Optimizing transformations for automated, high throughput analysis of flow cytometry data. BMC Bioinform 11:546-2105-11-546. doi:10.1186/1471-2105-11-546

Ge Y, Sealfon SC (2012) flowPeaks: a fast unsupervised clustering for flow cytometry data via K-means and density peak finding. Bioinformatics 28:2052–2058. doi:10.1093/bioinformatics/bts300

Gene Ontology Consortium (2004) The Gene Ontology (GO) database and informatics resource. Nucleic Acids Res 32:D258–D261

Gentleman RC, Carey VJ, Bates DM, Bolstad B, Dettling M, Dudoit S, Ellis B et al (2004) Bioconductor: open software development for computational biology and bioinformatics. Genom biol 5(10):R80

Hahne F, LeMeur N, Brinkman R, Ellis B, Haaland P, Sarkar D, Spidlen J, Strain E, Gentleman R (2009a) flowCore: a bioconductor package for high throughput flow cytometry. BMC Bioinformatics 10(1):106

Hahne F, Meur NL, Brinkman R, Ellis B, Haaland P, Sarkar D, Spidlen J, Strain E, Gentleman R (2009b) FlowCore: a bioconductor package for high throughput flow cytometry. BMC Bioinformatics 10:106

Herzenberg LA, Parks D, Sahaf B, Perez O, Roederer M, Herzenberg LA (2002) The history and future of the fluorescence activated cell sorter and flow cytometry: a view from Stanford. Clin Chem 48:1819–1827

Jelizarow M, Guillemot V, Tenenhaus A, Strimmer K, Boulesteix A (2010) Over-optimism in bioinformatics: an illustration. Bioinformatics 26:1990–1998

Kalina T, Flores-Montero J, van der Velden VH, Martin-Ayuso M, Bottcher S, Ritgen M, Almeida J, Lhermitte L, Asnafi V, Mendonca A, de Tute R, Cullen M, Sedek L, Vidriales MB, Perez JJ, te Marvelde JG, Mejstrikova E, Hrusak O, Szczepanski T, van Dongen JJ, Orfao A, EuroFlow Consortium (EU-FP6, LSHB-CT-2006-018708) (2012) EuroFlow standardization of flow cytometer instrument settings and immunophenotyping protocols. Leukemia 26:1986-2010. doi: 10.1038/leu.2012.122

Keeney M, Barnett D, Gratama J (2004) Impact of standardization on clinical cell analysis by flow cytometry. J Biol Regul Homeost Agents 18:305–312

Le Meur N (2013) Computational methods for evaluation of cell-based data assessment–Bioconductor. Curr Opin Biotechnol 24:105–111. doi:10.1016/j.copbio.2012.09.003

Levin E, Serrano K, Devine DV (2013) Biomedical Excellence for Safer Transfusion (BEST) Collaborative (2013) Standardization of CD62P measurement: results of an international comparative study, Vox Sang. doi: 10.1111/vox.12023

Lizard G (2007) Flow cytometry analyses and bioinformatics: interest in new softwares to optimize novel technologies and to favor the emergence of innovative concepts in cell research. Cytometry A 71:646–647. doi:10.1002/cyto.a.20444

Lo K, Brinkman RR, Gottardo R (2008) Automated gating of flow cytometry data via robust model-based clustering. Cytometry A. doi:10.1002/cyto.a.20531

Lugli E, Roederer M, Cossarizza A (2010) Data analysis in flow cytometry: the future just started. Cytometry A 77:705–713. doi:10.1002/cyto.a.20901

Maecker HT, McCoy JP, Nussenblatt R (2012) Standardizing immunophenotyping for the human immunology project. Nat Rev Immunol 12:191–200. doi:10.1038/nri3158

Maecker HT, Rinfret A, D'Souza P, Darden J, Roig E, Landry C, Hayes P, Birungi J, Anzala O, Garcia M, Harari A, Frank I, Baydo R, Baker M, Holbrook J, Ottinger J, Lamoreaux L, Epling CL, Sinclair E, Suni MA, Punt K, Calarota S, El-Bahi S, Alter G, Maila H, Kuta E, Cox J, Gray C, Altfeld M, Nougarede N, Boyer J, Tussey L, Tobery T, Bredt B, Roederer M, Koup R, Maino VC, Weinhold K, Pantaleo G, Gilmour J, Horton H, Sekaly RP (2005) Standardization of cytokine flow cytometry assays. BMC Immunol 6:13. doi: 10.1186/1471-2172-6-13

Maino VC, Maecker HT (2004) Cytokine flow cytometry: a multiparametric approach for assessing cellular immune responses to viral antigens. Clin Immunol 110:222–231. doi:10.1016/j.clim.2003.11.018

Overton WR (1988) Modified histogram subtraction technique for analysis of flow cytometry data. Cytometry 9:619–626

Pyne S, Hu X, Wang K, Rossin E, Lin TI, Maier LM, Baecher-Allan C, McLachlan GJ, Tamayo P, Hafler DA, De Jager PL, Mesirov JP (2009) Automated high-dimensional flow cytometric data analysis. Proc Natl Acad Sci U S A. doi:10.1073/pnas.0903028106

Reich M, Liefeld T, Gould J, Lerner J, Tamayo P, Mesirov JP (2006) GenePattern 2.0. Nat Genet 38:500–501. doi:10.1038/ng0506-500

Robinson JP, Rajwa B, Patsekin V, Davisson VJ (2012) Computational analysis of high-throughput flow cytometry data. Expert Opin Drug Discov 7:679–693. doi:10.1517/17460441.2012.693475

Roederer M (2001) Spectral compensation for flow cytometry: visualization artifacts, limitations, and caveats. Cytometry 45:194–205

Roederer M, Hardy RR (2001) Frequency difference gating: a multivariate method for identifying subsets that differ between samples. Cytometry 45:56–64. doi:10.1002/1097-0320(20010901)45:1 < 56:AID-CYTO1144 > 3.0.CO;2–9 [pii]

Roederer M, Moore W, Treister A, Hardy RR, Herzenberg LA (2001) Probability binning comparison: a metric for quantitating multivariate distribution differences. Cytometry 45:47–55. doi:10.1002/1097-0320(20010901) 45:1 < 47:AID-CYTO1143 > 3.0.CO;2-A [pii]

Sarkar D, Le Meur N, Gentleman R (2008) Using flowViz to Visualize Flow Cytometry Data. Bioinformatics

Spidlen J, Breuer K, Brinkman R (2012a) Preparing a Minimum Information about a Flow Cytometry Experiment (MIFlowCyt) compliant manuscript using the International Society for Advancement of Cytometry (ISAC) FCS file repository (FlowRepository.org). Curr Protoc Cytom Chapter 10:Unit 10.18. doi: 10.1002/0471142956.cy1018s61

Spidlen J, Breuer K, Rosenberg C, Kotecha N, Brinkman RR (2012b) FlowRepository: a resource of annotated flow cytometry datasets associated with peer-reviewed publications. Cytometry A 81:727–731. doi:10.1002/cyto.a.22106

Suni MA, Dunn HS, Orr PL, de Laat R, Sinclair E, Ghanekar SA, Bredt BM, Dunne JF, Maino VC, Maecker HT (2004) Performance of plate-based cytokine flow cytometry with automated data analysis. feedback

Zare H, Bashashati A, Kridel R, Aghaeepour N, Haffari G, Connors JM, Gascoyne RD, Gupta A, Brinkman RR, Weng AP (2012) Automated analysis of multidimensional flow cytometry data improves diagnostic accuracy between mantle cell lymphoma and small lymphocytic lymphoma. Am J Clin Pathol 137:75–85. doi: 10.1309/AJCPMMLQ67YOMGEW

# Shooting Movies of Signaling Network Dynamics with Multiparametric Cytometry

**Manfred Claassen**

**Abstract** Single-cell technologies like mass cytometry enable researchers to comprehensively monitor signaling network responses in the context of heterogeneous cell populations. Cell-to-cell variability, the possibly nonlinear topology of signaling processes, and the destructive nature of mass cytometry necessitate nontrivial computational approaches to reconstruct and sensibly describe signaling dynamics. Modeling of signaling states depends on a set of coherent examples, that is, a set of cell events representing the same cell state. This requirement is frequently compromized by process asynchrony phenomena or nonlinear process topologies. We discuss various computational deconvolution approaches to define molecular process coordinates and enable compilation of coherent data sets for cell state inference. In addition to the conceptual presentation of these approaches, we discuss the application of these methods to modeling of TRAIL-induced apoptosis. Due to their generic applicability these computational approaches will contribute to the elucidation of dynamic intracellular signaling networks in various settings. The resulting signaling maps constitute a promising source for novel interventions and are expected to be particularly valuable in clinical settings.

## Contents



M. Claassen (✉)
Institute for Molecular Systems Biology, ETH Zurich, Zurich, Switzerland
e-mail: mclaassen@ethz.ch

Current Topics in Microbiology and Immunology (2014) 377: 177–189
DOI: 10.1007/82_2013_350


Published Online: 6 August 2013

# 1 Introduction

Elucidation of the mechanisms used by cells to sense and process information is of fundamental scientific as well as clinical interest. This chapter aims to describe how to achieve this goal through use of multiparametric cytometry. The complexity of signaling systems poses a challenge for the elucidation of signaling mechanisms. Two types of complexity have a major impact on this process. First, signaling events typically happen in the context of heterogeneous cell populations (*intercellular complexity*), and, second, intracellular signaling systems are often comprized of many molecular components (*intracellular complexity*). These components span a heterogeneous spectrum of molecular entities: Kinases, phospho-sites, proteolytic cleavage sites, and conformational configurations. These components furthermore exhibit mutual interactions and possibly convoluted, noisy dependencies. These forms of complexity require experimental technologies that are able to generate informative data leading theoretical concepts that sensibly describe the dynamic and complicated succession of high-dimensional signaling states.

The introduction of genome-wide measurement techniques addressed the issue of *intracellular complexity* as they allow comprehensive monitoring of a signaling system with respect to all components/events of a certain class. Microarray and sequencing technologies allow mapping of transcriptional responses elicited by signaling system (Wang et al. 2009). These technologies are suited to monitoring of the late response to a signaling stimulus. In contrast, early signaling events typically involve components at the level of proteins and their post-translational modifications, for instance phosphorylations. Mass spectrometry-based proteomics approaches enable researchers to monitor this class of components on a proteome-wide scale (Altelaar et al. 2013). Mass spectrometry techniques can be used to analyze proteins, including a limited number of modification types, with great breadth across a proteome. This breadth comes at the cost of sensitivity. Contemporary mass spectrometry approaches are not routinely able to detect protein components in the sub-attomol range (Gillet et al. 2012; Picotti et al. 2009) and are of limited utility for monitoring of the behavior of a large proportion of low abundance, and yet important, signaling proteins. Furthermore, certain modification types, for instance conformational changes, are generally difficult to observe. Other modification types might not be observable for specific proteins, and these modifications may be relevant for the signaling system of interest.

Antibody-based detection strategies like the traditional western blot suffer from these sensitivity limitations to a lesser extent. The sensitivity of carefully chosen antibodies enables researchers to monitor very low abundance targets at levels close to single copies per cell. Antibodies can be raised to a wide range of epitopes, even those comprising unusual protein modifications or conformation-specific protein variants. These considerations exemplify how antibody-based strategies are suited to study very subtle mechanisms of signaling systems. However, such strategies can only be used to monitor components that are known a priori. For most fundamental signaling systems this is the case since their key components have been extensively studied.

The technologies discussed so far constitute solutions to specifically address the *intracellular complexity* of signaling systems. These population-based approaches all report average configurations across a large population of cells and cannot be used to resolve and might even mislead researchers due to *intercellular complexity* of signaling systems. Signaling events typically occur in the context of a heterogeneous cell population. Cell-type-specific signaling responses are likely to be averaged out in this situation if studied by means of population-based methods. Recent studies have shown that signaling processes exhibit asynchrony, that is, cell-to-cell variation of response rates in homogeneous cell populations (Spencer et al. 2009). These phenomena induce additional cell population heterogeneity over the course of signaling responses and therefore limit the applicability of population-based methods even in initially homogeneous model systems like cell lines.

Measurement technologies that operate on single cells have the potential to overcome the issues resulting from cell population heterogeneity. Such technologies achieve single-cell resolution by either resorting to microscopy or flow cytometry approaches. With exception of recent single-cell transcriptomics technologies (Kalisky et al. 2011), markers of interest have been measured on the basis of antibody labeling with histochemistry or fluorescence read-out. These methods benefit from the versatility and sensitivity of antibody-based strategies but are used to analyze individual cells. These methods have been limited by the number of markers that can be observed simultaneously, roughly a dozen for fluorescence-based technologies (Perfetto et al. 2004). This drawback has been largely alleviated with the introduction of the CyTOF, a highly multiparametric flow cytometry technique with mass spectrometry readout that allows quantitative monitoring of more than 30 molecules simultaneously at the single-cell level (Bandura et al. 2009; Bendall et al. 2011; Ornatsky et al. 2010). This situation enables for comprehensive systems-level analysis of signaling processes, successfully addressing both types of *intra-* as well as *intercellular complexity*.

Different computational approaches have been proposed to infer network models of the underlying signaling systems from the raw data produced by single-cell monitoring techniques. Network reconstruction typically starts from a set of snapshots acquired of a biological system under different conditions and aims to elucidate the molecular processes that gave rise to the data in first place. There have been many approaches to describe or model these processes. Mechanistic models describing each network relationship as chemical reaction are an appealing option due to their detail and physico-chemical justification (Szallasi et al. 2006). However, the scarcity of available data typically renders genome-wide application of these very complex models infeasible. Probabilistic modeling approaches seek to discover statistical dependencies in the data and are a flexible alternative since their complexity can be elegantly controlled (Tibshirani 1994). These approaches range from simple correlation (Rice et al. 2005) or regression analyses (Rogers and Girolami 2005) to Bayesian network modeling (Friedman 2004) to more structured and robust module network modeling (Segal et al. 2003) and extensions thereof (Lee et al. 2009).

This chapter will elaborate on the specific task of using perturbation time series experiments with mass cytometry readout to model signaling systems. This setup and computational analysis allows virtual filming of the response of systems to a perturbation in molecular detail. This review will describe the steps of the experiment and of the systems modeling. Specifically, we will describe the experimental setting and the technology underlying mass cytometry and discuss how to infer informative, static cell state representations from the resulting data and how to assemble these snapshots into movie-like representations of the studied signaling process.

## 2 Set and Actors: Time-Series Experiments of Signaling System Components

The study of signaling systems depends on our ability to monitor the behavior of relevant system components over time. Early signaling events typically occur on time scales that do not involve transcriptional regulation. Instead such events are governed by post-translational protein modifications like phosphorylations, methylations, ubiquitylations, glycosylations, proteolytic cleavages, or conformation changes (Haglund and Dikic 2005; Haines and Irvine 2003; Johnstone et al. 2008; Suganuma and Workman 2011).

Antibody-based detection strategies are well suited to detection of specific components that are indicative of these modifications. To fully benefit from the diverse specificity spectrum, it is necessary to know beforehand which components to target. For many fundamental signaling systems the set of components is established and therefore this requirement does not constitute a limitation for an antibody-based strategy. If the components of a network of interest are not known, an unbiased screen of an antibody library can specify the component list. This type of screen can be used to independently assess whether levels of particular library epitopes show significant variation across the studied process. If this screen is performed with a technology that allows single-cell resolution, like multiparametric cytometry, this assessment tests for cell-type-specific variation that might not be apparent from bulk measurements.

Signaling events occur in the context of a time course. For the following sections we therefore will assume an experimental setting in which, after initiating the signaling process of interest in a possibly heterogeneous cell population, a set of samples is collected at discrete time points and subjected to multiparametric cytometry. Initiation of the process typically involves the addition of an extracellular stimulus like a cytokine to activate a receptor-induced signaling cascade. The data generated for the samples collected over the time course serve as a basis for reconstruction of an informative model of the process of interest.

## 3 Camera: Mass Cytometer

Mass cytometry constitutes a form of multiparametric cytometry that is particularly suited to comprehensive study of signaling systems. These systems typically do not involve a set of components that spans a complete genome, instead up to several dozen relevant components are involved. With its potential to simultaneously detect over 30 markers, mass cytometry has the ability to measure all components of a signaling system at single-cell resolution.

Conceptually, mass cytometry borrows from conventional fluorescence-based flow cytometry. Cells are stained with reporter-conjugated antibodies with specificities for cell surface or intracellular targets and are analyzed on a cell by cell basis using a flow cytometer that is coupled to a device that is able to detect the reporter. The innovations for mass cytometry include: (1) The use of antibodies that are conjugated with stable isotopes of nonbiological, rare earth metals and (2) the readout of single-cell events in an inductively coupled plasma mass spectrometer (ICP-MS). Mass cytometry takes advantage of metal, rather than fluorescent, reporters. Mass spectrometer detection circumvents limitations related to spectral overlap and thereby enables to routine measurement of more than 30 different markers simultaneously (Bendall et al. 2012; Benoist and Hacohen 2011; Ornatsky et al. 2010). The scanning speed of the mass spectrometer allows analysis rates of up to 1000 cells per second. Multiplexing approaches further exploit the breadth and throughput of mass cytometry to simultaneously study the system of interest in up to 96 different conditions (Bodenmiller et al. 2012). The specifications of this high-content, single-cell analysis platform enable researchers to comprehensively and deeply study molecular systems that consist of dozens of molecular components. Due to the high throughput, mass cytometry experiments produce data on tens of thousands of cell events informing complex models that are able to describe very subtle relationships among molecular components.

## 4 Frames: Probabilistic Cell State Descriptions

Signaling processes are a temporal succession of cell states that can be thought of as atomic frames of a movie. This section will elaborate on various probabilistic representations of such cell states on the basis of multiparametric cytometry data. We will describe a cell state as a configuration of the set of molecular components for which we are acquiring data. The description of a cell state should, in addition to protein/modification abundances, ideally include information on the relationships between the monitored molecular players that for instance recapitulate pathway configurations and dependencies. Typically, relationships are not known. Furthermore, it is rarely possible to determine a suitable mechanistic model since model selection and parameter inference for this model class is rarely tractable in our high-dimensional setting (Wahl et al. 2006). Given a coherent set of cell events, probabilistic summaries are an appealing alternative to describe the underlying cell state without suffering from the latter limitations. Probabilistic approaches attempt to find and fit a suitable probability distribution to the given data. In our setting we are seeking to fit a multivariate density for the panel of marker abundances. The challenge of probabilistic modeling of cell states consists in confining the class of probability distributions such that a rich spectrum of dependencies can be discovered and at the same time model selection and parameter inference remain computationally tractable.

Probabilistic graphical models achieve a tradeoff between simplicity (i.e., few parameters and favorable properties for inference) and expressivity (i.e., a wide spectrum of correlation structures). Formally, these modeling approaches explicitly formulate dependence (or independence) assumptions among model variables (e.g., MAPK is conditionally dependent on MAPK levels). The multivariate probability distribution is typically visualized in a graph, in which variables are represented as nodes and dependence structure is encoded in the connecting edges. The dependence assumptions formally translate into factorizations of the respective probability distribution, in turn significantly simplifying mathematical operations for model selection and parameter inference (Bishop 2007). Two popular and well-understood classes of probabilistic models are Bayesian networks and Gaussian Markov random fields. Both have been extensively applied in computational biology (Friedman 2004) and other fields (Jordan 2004).

Bayesian networks resort to directed graphs to exemplify conditional dependencies among the variables. The connectivity structure of the graph defines conditional probability distributions $P(X \mid pa(X))$ for each variable $X$, where $pa(X)$ denotes the parent variables of $X$ (i.e., those variables with edges toward $X$). The joint probability distribution of all variables is represented as the product of conditional probabilities for each variable: $P(X_1, X_2, \ldots) = \Pi_i P(X_i \mid pa(X_i))$. Figure 1 provides an illustration of the probability decomposition induced in a Bayesian network. For a given data and a fixed noncyclic network structure, parameter estimation can be efficiently performed by belief propagation algorithms (Bishop 2007). For cyclic network structures, approximate inference procedures like loopy belief propagation are available. Although these methods lack theoretical optimality guarantees they have performed well in practice (Wainwright and Jordan 2008). Selecting the right network structure among the exponentially many possible ones constitutes a major computational bottleneck for Bayesian network modeling. Typically, one must solve the combinatorial problem of explicitly enumerating all or at least a significant fraction of different network topologies and performing parameter estimation on each of these (Koller and Friedman 2009). Bayesian network modeling has been applied to model cell states of primary immune system cells after various perturbations and conventional flow cytometry readout (Sachs et al. 2005). The graphical representation of the resulting models recapitulates known pathway relationships and suggested novel relationships in T cell signaling that could be validated by targeted follow-up experiments.

Markov random fields resort to an undirected graph representation of multivariate probability distributions. Maximally densely connected subgraphs $\boldsymbol{X_C}$

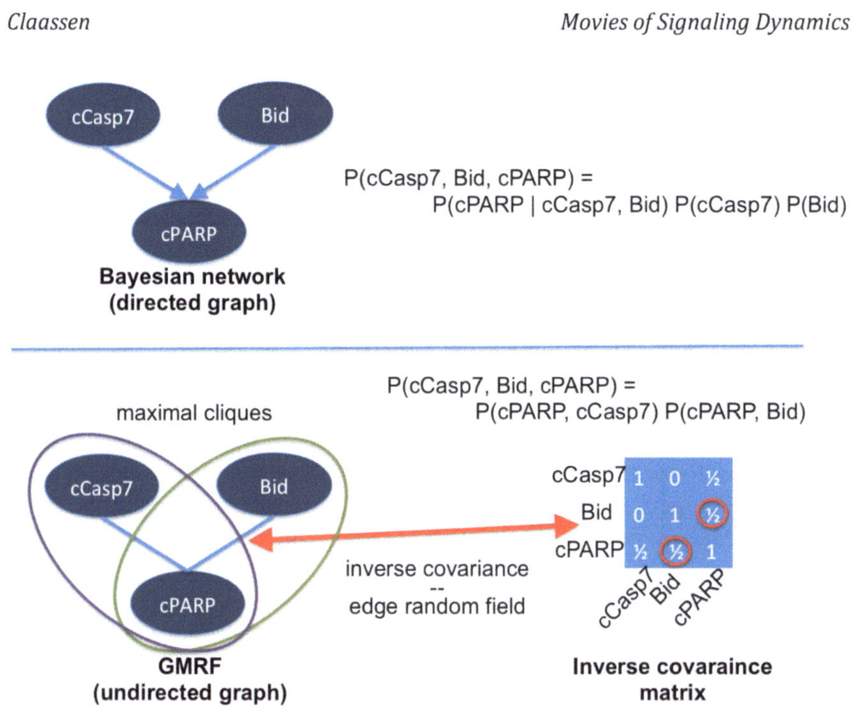


**Fig. 1** Illustration of a Bayesian network and a Gaussian Markov random field. The Bayesian network encodes the joint distribution over the three variables—cCasp7, Bid, and cPARP—as a product over the conditional distributions induced by the network connectivity. The Markov random field encodes the joint distribution as a product of the distributions over the maximal cliques of the graph. The Markov random field representation of a multivariate Gaussian distribution can be constructed directly from the inverse covariance matrix. Edges are drawn between nodes whose corresponding entries in the inverse covariance matrix are nonzero. Note that in this example, the Bayesian network and the Gaussian Markov random field do not represent the same joint probability distribution

(i.e., maximal cliques) define potentials $\psi_C$. The joint probability distribution of all variables is represented as the normalized product of the potentials: $P(X_1, X_2, \ldots) = \Pi_C\ \psi_C(X_C)$. In this formulation, two variables are conditionally independent if their respective nodes are not connected by an edge. Potential co-variation of independent variables is mediated by other system components that lie on the connecting path. Figure 1 illustrates the probability decomposition induced in a Markov random field.

To assess the potential of Markov random fields for cell state modeling we establish a very useful relationship between multivariate Gaussian distributions and their Markov random field representations. Multivariate Gaussian distributions have two parameters: The mean and the (inverse) covariance matrix. We are particularly interested in the (inverse) covariance matrix since it captures the relationship between variables. It can be shown that entries in the inverse covariance matrix can be directly translated into the underlying Markov random field (Koller and Friedman 2009). Specifically, there is an edge in the Markov random field if and only if the respective entry in the inverse covariance matrix is nonzero (Fig. 1). If data can be fit to a multivariate Gaussian then we can easily extract useful dependence/independence properties among the studied markers. In fact, it is computationally straightforward to fit the parameters of a multivariate Gaussian to a set of cell events (each represented with the log abundances of the measured markers). To perform model selection, that is to choose edges or the absence thereof in the underlying Markov random field, we have to choose non-zero or zero, respectively, entries in the inverse covariance matrix. This task can be elegantly accomplished by introducing sparsity-inducing priors to the parameter fitting procedure, resulting in an augmented continuous optimization problem (Friedman et al. 2008). This optimization problem is convex and can therefore be solved efficiently (Boyd et al. 2011). In summary, if we can assume that our (suitably transformed) data follows a multivariate Gaussian distribution, then we can efficiently perform model selection and parameter fitting to obtain the underlying Markov random field thereby revealing the topology of the signaling pathway.

We applied this strategy to describe cell states in the course of TRAIL-induced apoptosis in a colon cancer cell line (Claassen et al., manuscript in preparation). Preliminary empirical analysis of the logarithmically transformed data confirmed the validity of assuming a multivariate Gaussian distribution of the marker abundance. The inferred sparse inverse covariance matrices both recapitulated known relationships and revealed novel ones that were shown to be important events in decisions to fully commit to apoptosis or not.

This section reviewed different computational approaches to summarize high-dimensional signaling states and extract functionally relevant relationships among the monitored markers. These approaches yield a state model from a set of examples of the same signaling state. The dynamic nature of the studied processes frequently results in intricate phenomena that preclude the definition of homogeneous sets directly from the experimental data. The next chapter will discuss these phenomena and review computational approaches to define homogeneous sets of cell events in these situations.

## 5 Movie Plots: Asynchrony and Topology of Signaling Processes

Timing and topology of a signaling process have a strong impact on whether or not sensible cell state models can be inferred. It is necessary to provide cell events (examples) of a specific state to predict a model for that state. Unfortunately, a mass cytometry run rarely represents a homogeneous set of cell events for a single state. Instead, this set typically recapitulates a continuous spectrum of different states. Thus, homogeneous sets must be computationally defined on the basis of the experimental data in a first step before probabilistic models of individual signaling states can be deduced.

Signaling processes follow a trajectory in the high-dimensional space spanned by the markers monitored by mass cytometry. This trajectory can exhibit different types of topologies, for instance linear, bifurcated tree-like or cyclic topologies. Furthermore, considerable asynchrony (i.e., cell-to-cell variability for the traversal speed along the process trajectory) has been observed for certain processes like TRAIL-induced apoptosis (Spencer et al. 2009). These phenomena constitute the sources for signaling process-induced heterogeneity in cell populations studied by mass cytometry. Various approaches have been proposed to define molecular process coordinates that allow assignment of the degree of process accomplishment to individual cell events and thereby definition of sets of cell events with similar signaling states.

Asynchrony has been recently described for TRAIL-induced apoptosis (Spencer et al. 2009). This signaling process constitutes a receptor-induced variant of apoptosis and shares most components of the linear canonical apoptotic pathway (Johnstone et al. 2008). We observed asynchrony in TRAIL-induced apoptosis in a colon cancer cell line in a time-series experiment with discrete time point sampling and mass cytometry readout of 25 markers known to be important in TRAIL induced apoptosis. Due to asynchrony, a spectrum of different cell states is represented in each sample (Fig. 2). Due to this heterogeneity it is not sensible to use each sample in the cell state model for the respective time point. Instead, we sorted our cell events with respect to their progress in the studied process. Intracellular markers that from prior knowledge are indicative for the process progress guided this sorting procedure. For individual markers, monotonous increase/decrease over time was necessary to ensure a one-to-one mapping between marker abundance and process progress. In the context of TRAIL-induced apoptosis, it made sense to use pNFkB and cleaved caspase 3 as markers to resolve early and late signaling events, respectively. We used each of these markers sort cell events with respect to their apoptotic progression, regardless of the measurement time in the time-series experiments. Finally, we used a sliding window approach along the process coordinate marker to define a series of events in cells of similar signaling state (Fig. 2). These event sets were then used to develop a probabilistic state model as described in the previous section. By this means, we obtained a series of state models that recapitulated the known configuration changes along the progression of the studied process (Fig. 3). We also identified previously uncharacterized relationships so far unappreciated in the literature that are further studied in follow-up experiments at the moment. This result underscores the power of the mass cytometry approach to study signaling cascades. By jointly monitoring all relevant components of the signaling system, we can implicitly screen for all binary relationships regardless of our prior expectations and thereby facilitate the discovery of novel relationships.

The definition of process coordinate markers can be complicated by nonlinear topologies of the studied process. Different approaches have been proposed to

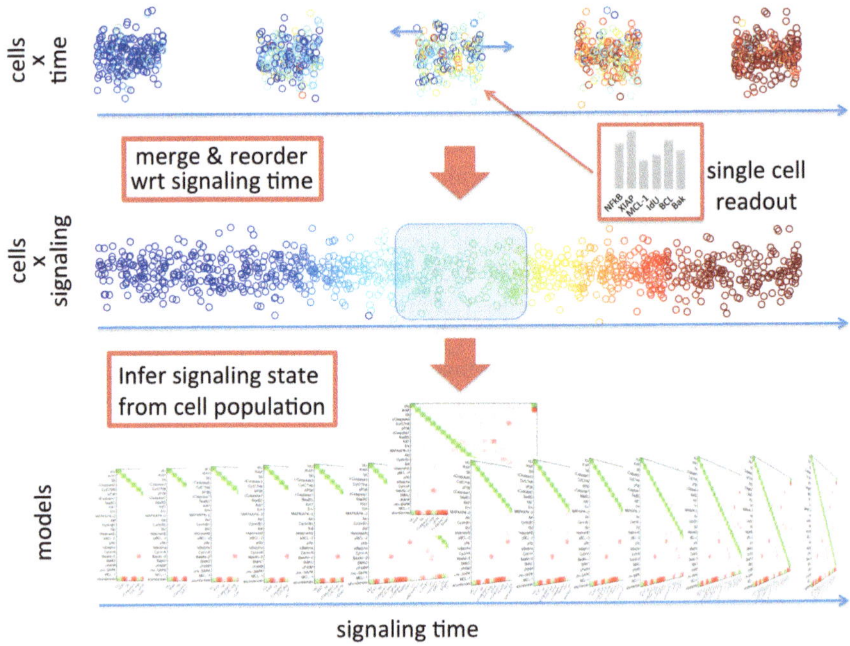


**Fig. 2** Shooting a movie of an asynchronous linear process. *Top* Illustration of the raw multiparametric data obtained from cell populations sampled at discrete time points. Each *circle* corresponds to a cell event. For each cell, data on 5–30 marker abundances are obtained depending on the type of cytometry. The coloring scheme (*blue* to *red*) encodes the degree of progression in the process (start to end). This scheme illustrates the asynchrony of the process (i.e., the cell-to-cell variability in the process rate) and indicates that the sampled cell populations are not homogeneous. *Middle* The introduction of a molecular process coordinate allows reconstruction of the degree of progression for each cell event and appropriate sorting. *Bottom* A sliding window approach can be used to define coherent subpopulations that can be used to infer models of the respective cell states. Here, we depict a consecutive series of inverse covariance matrices obtained from representing signaling states by Gaussian Markov random fields

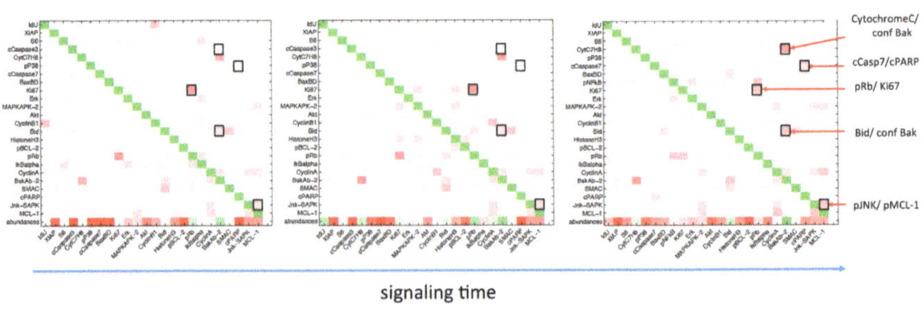


**Fig. 3** Three selected cell-state snapshots presented as inverse covariance matrices with relative marker abundances in *bottom row* for TRAIL-induced apoptosis in a colon cancer cell line. Twenty-five apoptosis markers were measured. Known relationships are highlighted

account for nonlinear topology. Chen et al. (manuscript in preparation) developed a computational approach to define a process coordinate for cyclic processes in the cell cycle as well as perturbations. Their approach relies on approximating the shortest Hamiltonian path in cluster of networks generated from flow cytometry data. Differentiation processes have a bifurcated tree topology and have been studied with mass cytometry. To define and characterize distinct cell populations along a differentiation tree, Qiu et al. proposed a spanning tree progression procedure called SPADE (Qiu et al. 2011). SPADE is a clustering procedure that explicitly takes into account that differentiation processes are continuous and that aligns cell populations along a tree-like trajectory. SPADE elegantly circumvents the difficult task of deciding on the number of clusters in an unsupervised learning setting. In a first step, the mass cytometry data is clustered into many small preliminary clusters. In a second step, these cluster centroids are connected by a minimum spanning tree. These methods can be used to specify the course of nonlinear trajectories of bifurcated or cyclic processes. The resulting trajectories can be used to define homogeneous cell populations in proximity to a position on this trajectory that correspond to a specific cellular state. Each of these populations can be described in depth by resorting to probabilistic models as described in the previous section.

## 6 Summary and Outlook

This chapter describes the use of multiparametric cytometry to describe the dynamic behavior of signaling systems. Multiparametric single-cell-resolving technologies can be used to elucidate these processes despite their intrinsic *intracellular* complexity and the complications related to their *intercellular* complexity. We have developed a computational approach that processes time-series experiments with multiparametric cytometry readout to generate a movie-like description of a signaling response (i.e., a series of consecutive signaling state snapshots). Probabilistic graphical models are suited to computationally summarize the raw data resulting from cytometric studies into interpretable and informative signaling state descriptions. Specifically, the application of Bayesian networks and Markov random fields has proven successful. Computational approaches can be used to infer high-dimensional process trajectories and molecular process coordinates. These approaches enable modelers to generate a series of consecutive signaling state snapshots for asynchronous processes with nontrivial topologies by allowing the definition of cell populations with a coherent signaling state. The conceptual computational approaches described were developed for analysis of cytometric data but are generalizable to other single-cell technologies like single-cell transcriptomics. It seems natural to adapt these ideas to these technologies to complement our models with respect to slow signaling-induced phenomena like transcriptional responses.

The movie-like description of a signaling process is informative as it reveals dynamically changing relationships that govern the signaling process but remains descriptive in nature. Addition of mechanistic details to this description would be an interesting and possibly insightful extension of this approach. This goal can be achieved serially by first learning the probabilistic description from data and align the relationships to a priori knowledge from the literature. An alternative approach would consist of formally introducing probabilistic relationships in a data-driven model selection procedure to find a suitable mechanistic model. We expect that highly informative multiparametric single-cell technologies in conjunction with the discussed generically applicable computational approaches will reveal novel mechanistic details about dynamic intracellular signaling networks in a variety of clinically relevant settings.

## References


Altelaar AF, Munoz J, Heck AJ (2013) Next-generation proteomics: towards an integrative view of proteome dynamics. Nat Rev Genet 14:35–48

Bandura DR et al (2009) Mass cytometry: technique for real time single cell multitarget immunoassay based on inductively coupled plasma time-of-flight mass spectrometry. Anal Chem 81:6813–6822

Bendall SC et al (2011) Single-cell mass cytometry of differential immune and drug responses across a human hematopoietic continuum. Science 332:687–696

Bendall SC et al (2012) A deep profiler's guide to cytometry. Trends Immunol 33:323–332

Benoist C, Hacohen N (2011) Immunology. Flow cytometry, amped up. Science 332:677–678

Bishop CM (2007) Pattern recognition and machine learning, vol 28. Springer, New York

Bodenmiller B et al (2012) Multiplexed mass cytometry profiling of cellular states perturbed by small-molecule regulators. Nat Biotechnol 30:858–867

Boyd S et al (2011) Distributed optimization and statistical learning via the alternating direction method of multipliers, vol 3. Now Publishers, Hanover

Friedman N (2004) Inferring cellular networks using probabilistic graphical models. Science 303:799–805

Friedman J, Hastie T, Tibshirani R (2008) Sparse inverse covariance estimation with the graphical lasso. Biostatistics 9:432–441

Gillet LC et al (2012) Targeted data extraction of the MS/MS spectra generated by data-independent acquisition: a new concept for consistent and accurate proteome analysis. Mol Cell Proteomics 11:O111 016717

Haglund K, Dikic I (2005) Ubiquitylation and cell signaling. EMBO J 24:3353–3359

Haines N, Irvine KD (2003) Glycosylation regulates Notch signalling. Nat Rev Mol Cell Biol 4:786–797

Johnstone RW, Frew AJ, Smyth MJ (2008) The TRAIL apoptotic pathway in cancer onset, progression and therapy. Nat Rev Cancer 8:782–798

Jordan MI (2004) Graphical models. Stat Sci 19:140–155

Kalisky T, Blainey P, Quake SR (2011) Genomic analysis at the single-cell level. Annu Rev Genet 45:431–445

Koller D, Friedman N (2009) Probabilistic graphical models: principles and techniques, vol 1., Adaptive computation and machine learning seriesThe MIT Press, Cambridge

Lee SI et al (2009) Learning a prior on regulatory potential from eQTL data. PLoS Genet 5:e1000358

Ornatsky O et al (2010) Highly multiparametric analysis by mass cytometry. J Immunol Methods 361:1–20
Perfetto SP, Chattopadhyay PK, Roederer M (2004) Seventeen-colour flow cytometry: unravelling the immune system. Nat Rev Immunol 4:648–655
Picotti P et al (2009) Full dynamic range proteome analysis of *S. cerevisiae* by targeted proteomics. Cell 138:795–806
Qiu P et al (2011) Extracting a cellular hierarchy from high-dimensional cytometry data with SPADE. Nat Biotechnol 29:886–891
Rice JJ, Tu Y, Stolovitzky G (2005) Reconstructing biological networks using conditional correlation analysis. Bioinformatics 21:765–773
Rogers S, Girolami M (2005) A Bayesian regression approach to the inference of regulatory networks from gene expression data. Bioinformatics 21:3131–3137
Sachs K et al (2005) Causal protein-signaling networks derived from multiparameter single-cell data. Science 308:523–529
Segal E et al (2003) Module networks: identifying regulatory modules and their condition-specific regulators from gene expression data. Nat Genet 34:166–176
Spencer SL et al (2009) Non-genetic origins of cell-to-cell variability in TRAIL-induced apoptosis. Nature 459:428–432
Suganuma T, Workman JL (2011) Signals and combinatorial functions of histone modifications. Annu Rev Biochem 80:473–499
Szallasi Z, Stelling J, Periwal V (2006) System modeling in cell biology: from concepts to nuts and bolts. The MIT Press, Cambridge
Tibshirani R (1994) Regression shrinkage and selection via the lasso. J R Stat Soc Ser B 58:267–288
Wahl SA et al (2006) Unravelling the regulatory structure of biochemical networks using stimulus response experiments and large-scale model selection. Syst Biol (Stevenage) 153:275–285
Wainwright MJ, Jordan MI (2008) Graphical models, exponential families, and variational inference. In: Foundations and trends in machine learning,. vol 1, 1st edn. Now Publishers, Hanover, pp 1 electronic text (305 p : ill) : digital file
Wang Z, Gerstein M, Snyder M (2009) RNA-Seq: a revolutionary tool for transcriptomics. Nat Rev Genet 10:57–63

# Hyperspectral Cytometry

**Gérald Grégori, Bartek Rajwa, Valery Patsekin, James Jones, Motohiro Furuki, Masanobu Yamamoto and J. Paul Robinson**

**Abstract** Hyperspectral cytometry is an emerging technology for single-cell analysis that combines ultrafast optical spectroscopy and flow cytometry. Spectral cytometry systems utilize diffraction gratings or prism-based monochromators to disperse fluorescence signals from multiple labels (organic dyes, nanoparticles, or fluorescent proteins) present in each analyzed bioparticle onto linear detector arrays such as multianode photomultipliers or charge-coupled device sensors. The resultant data, consisting of a series of spectral fingerprints characterizing every analyzed cell, are not compensated by employing the traditional cytometry approach, but rather are spectrally unmixed utilizing algorithms such as constrained Poisson regression or non-negative matrix factorization. Although

G. Grégori · V. Patsekin · J. Paul Robinson
Department of Basic Medical Sciences, College of Veterinary Medicine, Purdue University, West Lafayette, IN 47907, USA

J. Jones · J. Paul Robinson
Weldon School of Biomedical Engineering, Purdue University, West Lafayette, IN 47907, USA

B. Rajwa
Bindley Bioscience Center, Purdue University, West Lafayette, IN 47907, USA

G. Grégori
Mediterranean Institute of Oceanography, Aix-Marseille Université, Université du Sud-Toulon Var, CNRS, IRD, Marseille 13288, France

M. Furuki
Life Science Business Div, MBU, Sony Corporation, Tokyo 141-0001, Japan

M. Yamamoto
Phototek Laboratory Inc, Yokohama 220-0003, Japan

J. Paul Robinson (✉)
Purdue University Cytometry Laboratories, 1300 Lynn Hall, West Lafayette, IN 47907-2057, USA
e-mail: jpr@flowcyt.cyto.purdue.edu

Current Topics in Microbiology and Immunology (2014) 377: 191–210
DOI: 10.1007/82_2013_359


Published Online: 23 November 2013

implementations of spectral cytometry were envisioned as early as the 1980s, only recently has the development of highly sensitive photomultiplier tube arrays led to design and construction of functional prototypes and subsequently to introduction of commercially available systems. This chapter summarizes the historical efforts and work in the field of spectral cytometry performed at Purdue University Cytometry Laboratories and describes the technology developed by Sony Corporation that resulted in release of the first commercial spectral cytometry system—the Sony SP6800. A brief introduction to spectral data analysis is also provided, with emphasis on the differences between traditional polychromatic and spectral cytometry approaches.

## Contents



## 1 Spectral Analysis by Cytometry

Analytical flow cytometry (FC) is one of the most powerful single-cell analysis techniques available. Researchers in the life sciences, including basic cell biology, molecular biology and genetics, immunology, plant science, microbiology, environmental sciences, and oceanography, depend on FC to quantify cellular phenotypes and physiological responses of individual cells (Shapiro 2003). In order to perform quantitative measurements, FC relies on optical properties of biological particles (such as bacteria, algae, or mammalian cells) suspended in a liquid medium. On most currently available instruments, the forward-angle light scatter provides a proxy for the size of the particles, and the side-angle light scatter is used to characterize particle shape and internal structure. Both signals are created by interactions between individual bioparticles and a light beam produced by an external light source. Various fluorescence emission signals can be simultaneously collected following this excitation.

As the bioparticles flow rapidly through the instrument detection chamber (often called a flow cell) at rates of up to several thousand particles per second, the fluorescence emission data are automatically collected using photomultiplier tube (PMT) arrays, quantified and digitized via fast electronics, and eventually stored on a computer for further statistical analysis. These data are subsequently processed by an operator using one of several dedicated software packages capable of visualizing and discriminating various populations of particles according to their optical properties. Cells or particles can be classified based on morphology, abundance of fluorescence labels, physiology, functional activity, or expression of certain cell-surface or internal antigenic determinants. The overall aim of cytometry analysis is to characterize heterogeneous cellular populations by decomposing them into a set of phenotypically different, but internally similar, groups described by biological function.

Although FC is perhaps the most widely used method for phenotypic analysis and classification, the core technology has remained essentially unchanged until recently. In almost all the current commercial FC systems, scatter and fluorescence signals travel down an optical pathway through a set of dichroic filters, each of which splits the incoming signal into two directions according to the wavelength bands selected. A signal from each fluorochrome is redirected in this manner until it reaches a dedicated point of acquisition where it is filtered through a specific band-pass filter of a desired wavelength before being collected on a dedicated photodetector that provides a current output proportional to light intensity. The charge or the current signal is subsequently converted into a voltage that can be readily digitized by an analog-to-digital converter and finally is recorded by a computer. In most commercial instruments, photodiodes detect the bright forward light-scatter signals, and photomultiplier tubes are used to collect the weak fluorescence emission signals; the latter typically require amplification (alternative single-cell technologies that are not dependent upon fluorescence, such as mass cytometry, are discussed elsewhere in this volume).

Over the past three decades, FC technology has developed from single-color (single band, single-fluorescence intensity) measurement systems through two-, three-, and four-color instruments to the newest benchtop instruments with 10 to 12 fluorescence detection channels. Although a collection of as many as 17 simultaneous separate fluorescence signals has been reported in a traditional cytometry experiment, typical FC systems collect a more manageable number of bands, typically between 5 to 10 (Perfetto et al. 2004; Roederer et al. 1997; De Rosa et al. 2001; Wang et al. 2009).

Despite the enormous progress in multiband (also called "polychromatic") cytometry, it has been recognized for many years that collection of the full emission spectra would provide significantly more information than measurement of just few predefined bands. This type of collection would also allow for more flexible instrument design. In 1979, Wade et al. reported the fluorescence spectrum recorded for particles in a flow system (Wade et al. 1979); however, the instrumentation recorded only integrated spectra from a large number of particles. Since data collection was not achieved at the individual particle level, the sample was obviously assumed to be homogeneous. In 1986, Steen and Stokke were able to measure averaged fluorescence spectra of rat thymocytes. They used a custom-built cytometer equipped with a grating monochromator (Steen 1986; Stokke and Steen 1986). In 1990, Buican proposed the use of a Fourier-transform interferometer to collect single-cell spectra, but in practical performance, the design had severe limitations as the cells had to remain in the laser beam for a relatively long time in order to obtain a measurable signal set (Buican 1990). By comparison, the time period available for the laser excitation on current high-speed FC systems lasts only from 1 to 10 microseconds. In 1996, Gauci et al. described a system based on a flint-glass prism and an intensified photodiode array (Gauci et al. 1996). Again, the low data acquisition rate precluded practical use, and, in addition, the efficiency of the photodiodes was inadequate. The same year Asbury et al. reported measurement of spectra of cells and chromosomes using a monochromator (Asbury et al. 1996). However, the design required that the wavelength be changed during the course of measurement, making continuous flow measurements impossible (Asbury et al. 1996; Gauci et al. 1996). The technique was limited to measurement of just a single band of fluorescence from any single particle. Other groups, including those from SoftRay Inc. and the Universities of Wyoming and Utah, pursued another prism-based concept, but no subsequent data were reported (Johnson et al. 2001).

## 2 Modern Hyperspectral Cytometry

The introduction of multianode photomultipliers by Hamamatsu (H7260 series) revigorated the attempts to construct a hyperspectral flow cytometer able to collect an approximation of a full spectrum from every single bioparticle in flow. In the early 2000s, researchers at Purdue University Cytometry Laboratories began work on hardware and software prototypes for fast classification of hydrodynamically focused bioparticles using a spectral detector extension attached to a commercial FC system (an EPICS Elite cell sorter, from Beckman Coulter). The concept assumed utilization of the recently available first-generation, 32-channel multianode PMT (Hamamatsu) that had also been used in the field of confocal microscopy (Robinson et al. 2007). The design rationale was to reduce the complexity of FC optical pathways by reducing the number of elements and replacing them with a single multiband detector capable of providing sufficient sensitivity, portability, and robustness (Grégori et al. 2012). The preliminary results documenting the early work on the multispectral FC were presented in 2004.[1] The data demonstrated that the technology had potential for future commercialization (Robinson 2004). As second-generation multianode PMTs offered better sensitivity, the second prototype was capable of simultaneously collecting 32 bands of fluorescence from each flowing particle in less than 5 μs (Robinson et al. 2005). This created new opportunities for analysis and characterization of cells in a high-throughput and high-content setting, but it also required advanced control software capable of handling increased numbers of parameters.

[1] Presented at the International Congress of Analytical Cytology, May 23–28, 2004, Montpellier, France.

The work on spectral cytometry was also progressing in other laboratories. In 2006, Goddard et al. presented an alternative design concept employing a diffraction grating and a charge-coupled device (CCD) detector. This device dispersed the collected signals (fluorescence and side-scattered light) onto a CCD image sensor coupled to a spectrograph (Goddard et al. 2006). The design of the instrumentation involved minimal modifications around the flow chamber and collection optics of a conventional flow cytometer. Unfortunately, the flow rate was highly restricted owing to limited sensitivity of the CCD. Recently, John Nolan and collaborators demonstrated a new-generation, CCD-based spectral cytometry system. In this implementation, a broadband volume-phase holographic grating is interfaced with an electron-multiplying CCD detector. The system offers spectral resolution of approximately 11 nm (Nolan et al. 2013).

Although most of the efforts in spectral cytometry development remained focused on fluorescence, in 2008 John Nolan's group at La Jolla Bioengineering Institute presented a Raman spectral flow cytometry concept (RSFC) in which surface-enhanced Raman detection (SERS) and flow cytometry were combined (Watson et al. 2008). The CCD-based detector on the system was sensitive enough to detect SERS spectra in samples containing nanoparticle tags bound to microbeads. It was also capable of measuring Raman spectra from particles bearing as few as 200 Raman tags and had an integration time as short as 100 microseconds. Results obtained with the instrument indicated that it could detect more probes in the spectral range used than traditional fluorescence-based systems, thus offering a powerful tool for signal multiplexing. The development of robust tags remains an important challenge, however, as nanoparticle-based SERS labels tend to be relatively heterogeneous compared to organic fluorochromes or fluorescent proteins, which can be prepared with higher purity. Even though researchers have made SERS systems reproducible, the processes still require significant development (Brown and Doorn 2008a, b).

Although the well-established cytometry vendors largely ignored the new technology, Sony Corp.—a relative new comer in the field of cytometry—pursued an advanced hyperspectral design of their own. The Sony concept utilized a multianode PMT and featured a very complex prism-based monochromator. Sony demonstrated a prototype instrument and reported on hyperspectral technology during the ISAC congress in Seattle in 2010[2] and announced the launch of the new hyperspectral flow cytometer product—a SP6800 Spectral Cell Analyzer—in 2012. Although most of the mentioned implementations provided valuable scientific contributions, only the design using multiarray PMT technology has impacted the development of new generations of commercial instruments, as exemplified by the multispectral cytometry system designed by Sony.

---

[2] Motohiro Furuki, Shingo Imanishi, Masaya Kakuta, Masanobu Yamamoto, Yohei Morita,Yuji Yamazaki, Yumiko Ishii, Hiromitsu Nakauchi, "Hyper-Spectral Flow Cytometer with a Microfluidic Chip." Presented at CYTO2010, XXV Congress of International Society for Advancement of Cytometry, May 9–12, 2010, Seattle, WA, USA.

## 3 Practical Issues of Hyperspectral Cytometry

The technical aspects of hyperspectral flow cytometry are often discussed and compared with analogous spectral imaging techniques. Indeed, in the fields of high-resolution optical microscopy and small-animal imaging, one of the most significant recent technical developments has been the commercialization and wide acceptance of spectral imaging approaches (Zimmermann et al. 2003). However, the fact remains that the tools and techniques developed in spectral imaging are not readily transferable to the realm of FC owing to the following issues:

- In flow cytometry, the time for data collection is in the microsecond range. The particles or cells pass through a liquid-handling system and hydrodynamic forces within the flow chamber result in a single, central core stream. Once this hydrodynamic focusing has been accomplished, the particles (cells) usually pass very quickly (within a few microseconds) through a very narrow and focused beam of intense laser light, during which time a large number of variables—such as light scatter and spectral signatures—are collected and recorded. This high speed of flow is the core feature of cytometry that allows for the collection of data on several thousand particles per second.
- Separation of all the optical signals must be achieved within the time scale of the measurement system (i.e., a few microseconds), which eliminates all the tunable filter approaches.
- Significantly, many more fluorescence labels are used simultaneously in flow cytometry experiments than in typical imaging-based cytometry measurements.
- Cells cannot be analyzed more than once in regular FC systems (i.e., one cannot average multiple measurements from a single cell). Thus, one must collect all required data in a single measurement cycle. In imaging systems it is possible to scan, rescan, and average up to the photobleaching limits.
- In flow cytometry, every particle (cell) is a distinctive entity; FC allows true single-cell analysis. Each signal from every cell in a population is considered unique; therefore, it is not acceptable to average signals from multiple cells, and it is not possible to achieve the same population classification results if one does so.
- Cytometry is inherently quantitative. Therefore reproducibility and standardization issues are fundamental. On the other hand, spectral analysis methods in imaging are often used to provide qualitative results.

Most previous attempts at implementing spectral detection in flow cytometry failed to result in an instrument that could perform multicolor measurements with the required sensitivity and speed. The family of Hamamatsu's multiarray PMTs used in the prototype developed at Purdue as well as in a commercial

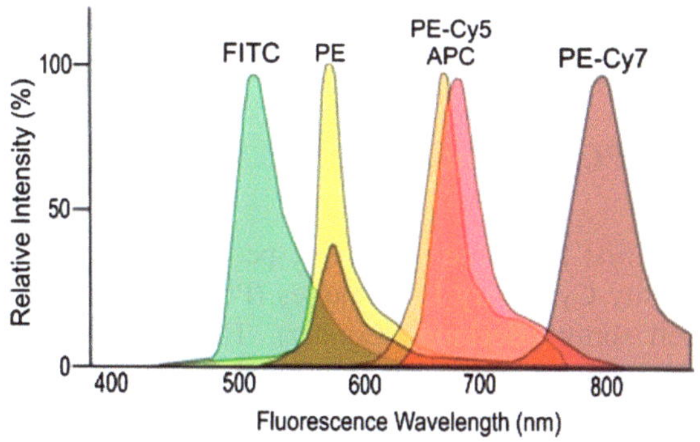


**Fig. 1** Typical extent of overlap of fluorescence for various molecules currently used in flow cytometry

implementation build by Sony (in a heavily modified version) are the first photodetectors that could be employed in multispectral cytometry setting.

## 4 Basic Analytical Problems to be Addressed

Both the absorption and emission spectra of the fluorochromes used in FC may carry valuable spectral information about tagged biological particles. The commonly used optical design of FC instruments requires that researchers employ a series of fluorochromes that have narrow excitation maxima and produce reasonably narrow emission bands within the sensitivity of the detector.

To achieve multiplexing and perform experiments with several fluorescent probes, a variety of excitation sources (multiple lasers offering multiple wavelengths) and a well-designed panel of fluorescence probes with minimally overlapping emission spectra must be used. When combining several labels in a sample, however, overlap of the emissions always occurs to some extent (Fig. 1). Indeed, most currently used fluorescent probes are organic and tend to have rather broad excitation and emission spectra. Therefore, it is virtually impossible to measure a signal from one fluorochrome while completely excluding the emission from all the others. The experimental setup, as well as the proper linear unmixing of the signals (compensation), can be very difficult, as exemplified by the complicated configurations of systems employed to perform highly polychromatic cytometry analysis (Perfetto et al. 2004; Roederer 2001; Roederer et al. 1997; De Rosa et al. 2001). As researchers attempt to use more fluorochromes simultaneously, current techniques utilizing multiple individual PMTs are reaching their limit and are becoming extremely complex, expensive, and difficult to scale further. To harness the full potential of new probes, such as nanocrystals, and to perform multiparametric experiments easily and effectively, alternative technologies must be considered (Bernstein and Hyun 2012; Nolan et al. 2013).

# 5 Hyperspectral Cytometry Instrumentation and Data Acquisition

## 5.1 PUCL Prototype

The first prototype of a hyperspectral flow cytometer developed at Purdue University Cytometry Laboratories (PUCL) utilized a heavily modified EPICS Elite cell sorter (Beckman Coulter). The data collection unit was comprised of a traditional custom-built polychromatic detection system using dichroic filters and a hyperspectral subunit designed at PUCL. The traditional detection module was based on a 30-nm FWHM polychromator from Asahi Spectra USA, Inc. This unit was equipped with six photodetectors (PMTs) and a set of band-pass filters (525/30 nm, optimized for FITC; 575/30 nm, PE; 620/30 nm, PE-Texas Red; 675/30 nm, PE-Cy5; 767/30 nm, PE-Cy7). To split the fluorescent signal coming from the measured particles in order to perform simultaneous measurement using the two units (the 32-channel PMT and the six-channel detection module), a 50/50 beam splitter was placed between the 32-channel spectral subsystem and the six-channel device. The EPICS Elite cell sorter was interfaced with these two devices to allow for simultaneous spectral and polychromatic data collection. Equivalent cell classification results were obtained using the significantly simplified spectral optical path and traditional detection components.

The spectral subsystem was comprised of a phase-volume holographic grating (Kaiser Optical Systems), which dispersed the signal onto a Hamamatsu 7260-01 32-channel multianode PMT. This multianode linear array offered a cathode sensitivity of 250 μA/lm and high uniformity between each anode (Grégori et al. 2012). The data acquisition was performed by a custom-developed software package (Fig. 2) equipped with all the conventional flow cytometry analysis features (histograms, scatter plots, region gating, back-gating, basic statistics) as well as statistical processing for spectral data analysis (including principal component analysis and conversion of data vectors into hyperspherical coordinates).

The second Purdue prototype, shown in Fig. 3, was developed using a modified FC500 flow cytometer (Beckman Coulter). The spectral detection module utilized a custom-enhanced version of Hamamatsu's high sensitive compact spectrometer (HSS A10766). The data acquisition system was upgraded to Beckman Coulter Gallios electronics capable of handling multichannel data collection. Experiments using fluorescent microspheres and lymphocytes labeled with a cocktail of antibodies (CD45/FITC, CD4/PE, CD8/ECD, CD3/Cy5) demonstrated the ability of the prototype to simultaneously collect 32 narrow bands of fluorescence from single particles flowing at about 1,000 events/second across the laser beam. The 32 discrete values collected at the single-cell level provide a proxy of the full fluorescence emission spectrum measured for each particle (Fig. 4).

Obviously, as the number of collected variables has increased, an advanced statistical processing is required in order to separate various clusters of cells analyzed in the spectral system (Fig. 5). The data analysis can be performed

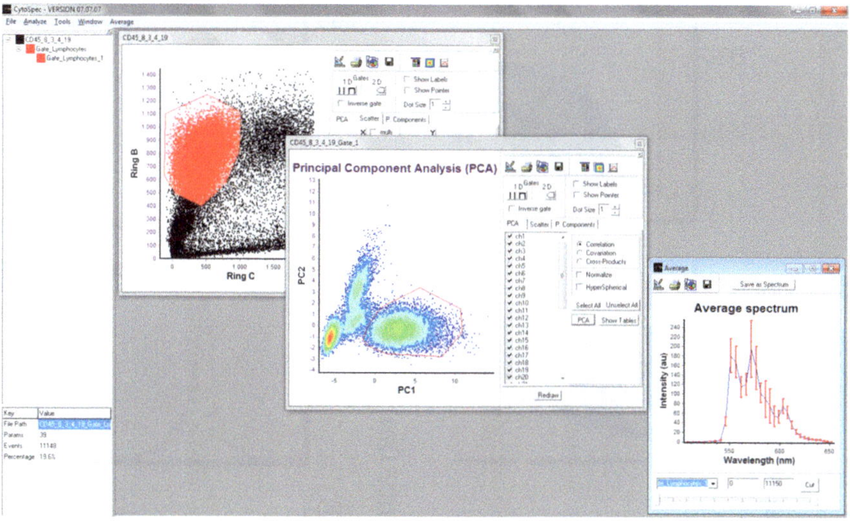


**Fig. 2** Screenshot of cytospec software package capable of acquisition and processing of hyperspectral cytometry data. The package was developed at PUCL by Valery Patsekin and is freely available at http://www.cyto.purdue.edu/Purdue_software

**Fig. 3** Photograph of a modified FC500 cytometer (Beckman Coulter) equipped with data acquisition subsystem from a Gallios instrument (Beckman Coulter) and the hyperspectral module developed at PUCL

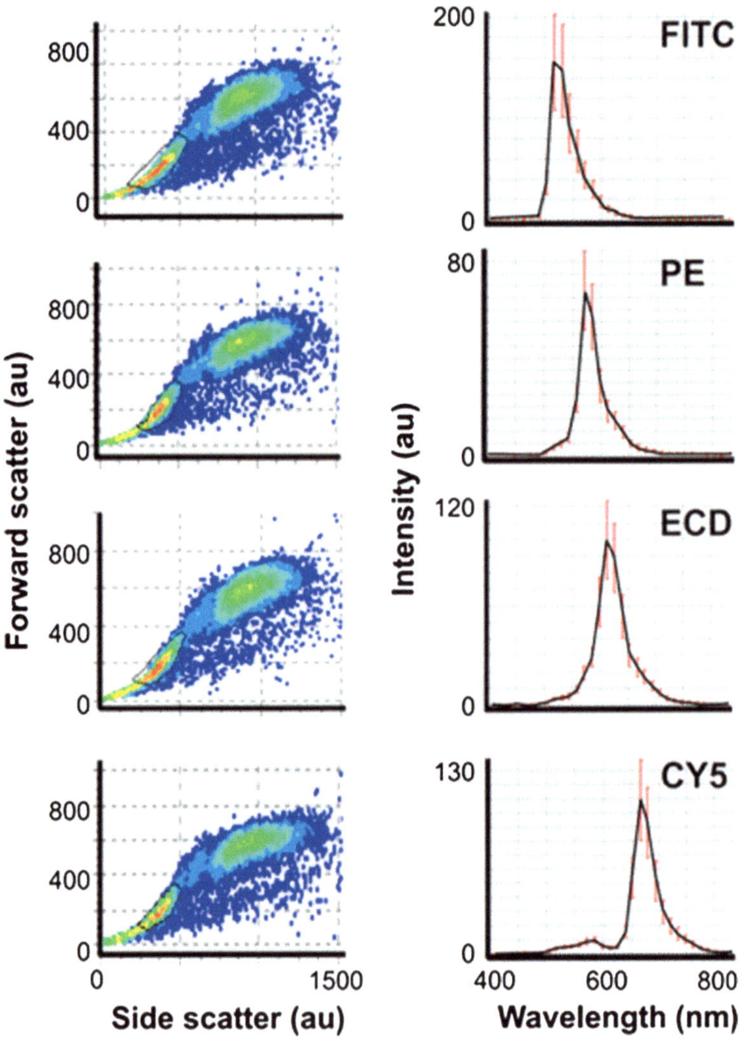


**Fig. 4** *Left column*, forward versus side-scatter cytogram recorded for control samples consisting of blood labeled with a single antibody conjugated with FITC, PE, ECD, or Cy5. *Right column*, corresponding average spectrum (and standard deviation for each channel) obtained for each single-stained control from the spectra of ~7,000 lymphocytes

employing either linear unmixing techniques followed by gating or dimensionality reduction approaches paired with supervised classification (Grégori et al. 2012; Novo et al. 2013).

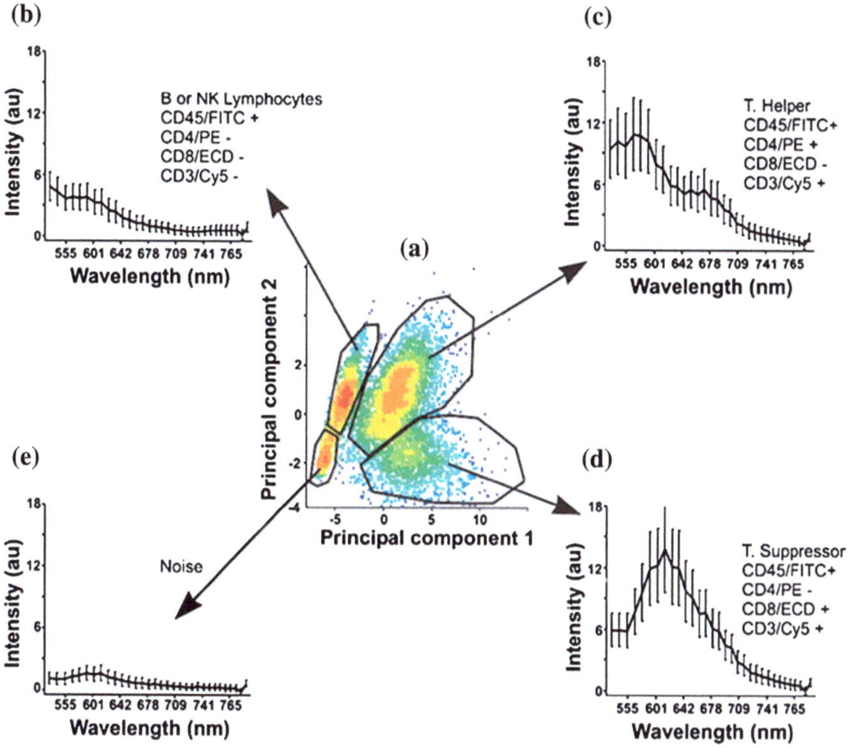


**Fig. 5** Different subpopulations of blood cells are clustered and classified on the basis of spectral data without the need for unmixing. Result of a principal component analysis performed on a flow cytometry data file corresponding to a blood sample incubated with a cocktail of antibodies labeled with four fluorochromes (FITC, PE, ECD, and Cy5). **a** Lymphocytes were first gated out from the rest of the particles on a forward-scatter versus side-scatter cytogram. In this example, four groups were discriminated based on the principal component analysis. The average spectrum (and the standard deviation per channel) is displayed for **b** B or natural killer lymphocytes, **c** T helper lymphocytes, **d** T suppressor lymphocytes, and **e** noise/detritus

## 5.2 Sony SP6800 Spectral Analyzer

The recently introduced Sony spectral analyzer offers a unique detection unit that captures all the emitted sample fluorescence ranging from 500 to 800 nm. This is achieved through a novel optical path utilizing a system of ten consecutive prisms and a custom-built 32-channel PMT, with independent voltage-controllable anodes (Fig. 6). Sample emission is directed through the prism set in order to disperse the signal into multiple separate bands (colors). A custom microlens array assembly then focuses each band of light onto a specific channel of the PMT array. This design limits the photon loss due to the presence of plates dividing the PMT channels and minimizes the crosstalk between the adjacent channels. The system is

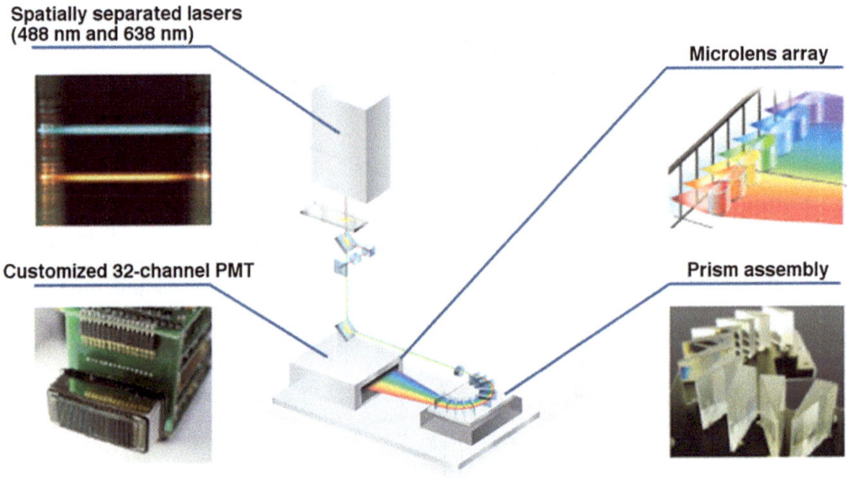


**Fig. 6** Schematic of Sony SP6800 spectral cell analyzer. Spatially separated lasers (488 nm and 638 nm) excite the labeled cells flowing through the flow chamber chip. The fluorescence spectra of the sample in the range of 500–800 nm are collected by a 32-channel linear array PMT detector equipped with a multi-prism monochromator

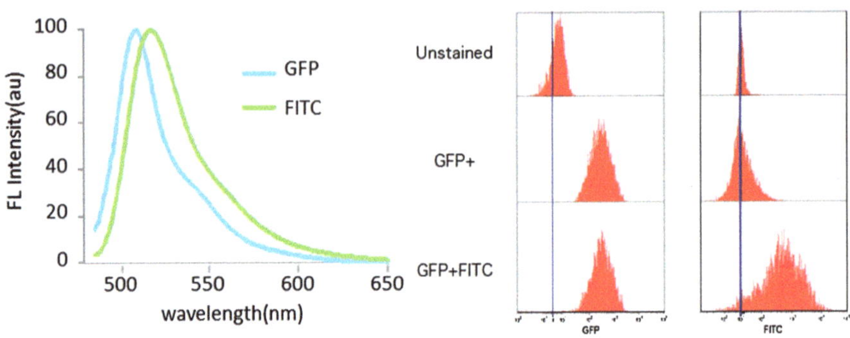


**Fig. 7** Separation of GFP and FITC signals using Sony spectral system. *Data* courtesy of W.C. Hyun (UCSF)

capable of performing up to 15-color analyses with only two lasers (488 nm and 638 nm). With this functionality, the SP6800 has the ability to determine the spectral profile of autofluorescence from a cell and automatically remove it from a stained sample, improving signal-to-noise and data accuracy. Figure 7 shows an example of simultaneous detection and unmixing of spectra derived from adjacent fluorescent proteins and fluorochromes (such as GFP and FITC), which traditional flow cytometers cannot separate.

In addition, the SP6800 is capable of combing the channels in the PMT to form "virtual filters" or "virtual bands." This sophisticated arrangement eliminates the need to use a secondary detection system of optical filter assemblies that are employed in standard flow cytometers. This also allows the SP6800 to function as a regular polychromatic, multiparametric FC with the capability of fine-tuning the detection to adjust for particular staining strategies and for problems with label intensities or marker abundance. Another unique function, adapted from Sony's DVD and Blu-ray laser-tracking technology, allows the researcher to gather information about each sample component by determining its location within the flow cell. This, in turn, can be used in cytometry analysis to decrease signal variability and improve coefficients of variations for the whole sample or individual populations as are assessed during cell cycle analysis, for example.

## 6 Spectral Data Analysis

In addition to differences in hardware, a crucial difference between spectral cytometry and multiparametric polychromatic cytometry is in the approach to data processing, specifically signal unmixing. Traditional cytometry relies on a process known as compensation for correction of spectral overlap. As mentioned before, the wide collection bands of conventional cytometers and the broad emission spectra of organic fluorochromes lead to significant spectral overlap between the signals emitted by the various fluorochromes. The currently used compensation approach simply implements in a software format the hardware-based concept proposed by Parks in 1977 (Loken et al. 1977). In the simplest two-band setting the compensation circuitry is comprised of two differential amplifiers. Since one signal is routed through a potentiometer to the positive side of one amplifier and the negative side of the other, a fraction of one signal can be subtracted from the other signal and vice versa. The idea can be expressed mathematically as

$$\begin{cases} a_1 = r_1 - qa_2 \\ a_2 = r_2 - pa_1 \end{cases} \Rightarrow \begin{cases} r_1 = a_1 + qa_2 \\ r_2 = pa_1 + a_2 \end{cases}$$

where $a_1$ and $a_2$ are the abundances of fluorochromes 1 and 2, parameters $q$ and $p$ describe the proportion of the fluorochromes signal measured in "incorrect" detectors owing to spectral overlap, and $r_1$ and $r_2$ are the actual measurements of fluorochrome intensities.

In the vector/matrix convention the assumed process of signal formation and the subsequent compensation can be summarized as

$$[r_1 \quad r_2] = \begin{bmatrix} 1 & q \\ p & 1 \end{bmatrix} [a_1 \quad a_2] \quad \Rightarrow \quad [a_1 \quad a_2] = \begin{bmatrix} 1 & q \\ p & 1 \end{bmatrix}^{-1} [r_1 \quad r_2]$$

The matrix $\begin{bmatrix} 1 & q \\ p & 1 \end{bmatrix}$ describing the proportion of fluorescence intensity that is measured in a channel other than the channel dedicated to a particular fluorochrome is called the spillover matrix (denoted herein by **S**). The inversion of this matrix is known as the compensation matrix **C**. The "compensated" result can be easily found by multiplying the measured signal by the compensation matrix:

$$[a_1 \quad a_2] = \mathbf{S}^{-1}[r_1 \quad r_2] = \mathbf{C}[r_1 \quad r_2] = \left[\tfrac{r_1 - qr_2}{1-pq} \quad \tfrac{pr_1 - r_2}{1-pq}\right]$$

Obviously, the compensation process can be generalized to any number of fluorochrome detectors (Bagwell and Adams 1993). In the matrix notation:

$$\begin{aligned} \mathbf{r} &\approx \mathbf{Sa} + \mathbf{n} \\ \mathbf{a} &\approx \mathbf{S}^{-1}\mathbf{r} = \mathbf{Cr} \end{aligned} \tag{1}$$

where **r** denotes the vector of observations of length $L$ (the number of detector channels/bands employed in the hyperspectral FC system), **S** an $L \times f$ spectral-spillover matrix ($f$ being the number of labels used in an experiment), **a** the vector of length $f$ of fluorochrome abundances, and **n** a vector of length $L$ that denotes noise.

As illustrated above, the "compensated" signal (that is, a signal with a fraction of unwanted signals removed) can easily be found by inverting a spillover matrix, which is essentially a matrix representing spectra of all used fluorochromes normalized to the peaks (hence, one on the matrix diagonal). Owing to the spectral overlap, the broad spectra of the typical organic fluorochromes used as labels, and the increased number of spectral bands, the compensated signal becomes relatively smaller and smaller. The signal-to-noise ratio in the individual bands decreases as well.

The rationale behind compensation assumes two important arrangements in the experimental setting. Firstly, the number of fluorochromes is identical to the number of spectral channels (detectors) used in the experiment. A single band/detector is dedicated to measurement of signal from a single fluorochrome. In other words, the signal from a fluorochrome spilling over to other channels is treated as unwanted background and removed.

Hyperspectral cytometry differs substantially in the experimental arrangement, and consequently in the assumptions regarding signal formation, data collection, and signal unmixing. In hyperspectral cytometry data analysis, fluorescent labels are not considered to be necessarily linked with single dedicated detection bands. Every label emits a spectrum, which potentially can be detected in all the bands. The resultant signal measured from cells labeled with multiple fluorescent tags is a linear combination of these spectra:

$$\mathbf{r} = \mathbf{Ma} + \mathbf{n} \tag{2}$$

where **M** is a spectral signature matrix (or mixing matrix), in which columns denote spectra of the fluorochromes used, and **a** and **n** are, as previously defined, related to fluorochrome abundance and noise, respectively.

This reasoning can be applied to flow cytometry data regardless of the number of spectral bands, and it follows a long—established methodology of spectral unmixing. The simplest version of the concept is well illustrated in the manuscript by Bagwell and Adams (Bagwell and Adams 1993), as well as in many spectral imaging publications (Garini et al. 2006; Keshava and Mustard 2002; Nielsen 2001; Settle and Drake 1993; Zimmermann 2005). The spectral unmixing (dubbed "additive compensation" in the cytometry field) involves addition of all portions of unmixed signal originating from a given fluorochrome. Mathematically it uses a notation very similar to that of compensation:

$$\begin{cases} r_1 = pa_1 + (1-q)a_2 \\ r_2 = (1-p)a_1 + qa_2 \end{cases}$$

$$[r_1\ r_2] = \begin{bmatrix} p & 1-q \\ 1-p & q \end{bmatrix} [a_1 \quad a_2]$$

One may notice that the mixing matrix $\mathbf{M}$ is columnwise normalized to 1. Consequently, even for a two-color arrangement, the spectral unmixing approach leads to a different result than does compensation:

$$\begin{aligned} [a_1 - a_2] &= \begin{bmatrix} p & 1-q \\ 1-p & q \end{bmatrix}^{-1} [r_1 - r_2] = \mathbf{M}^{-1}[r_1 - r_2] \\ &= \begin{bmatrix} \frac{qr_1+(q-1)r_2}{p+q-1} & \frac{pr_2+(p-1)r_1}{P+q-1} \end{bmatrix} \end{aligned}$$

If the number of spectral bands is equal to the number of fluorochromes, the general solution can be expressed as

$$\hat{\mathbf{a}} = \mathbf{M}^{-1}\mathbf{r}$$

where $\hat{\mathbf{a}}$ is an estimation of pure fluorochrome intensities, and hence their abundances. The number of spectral bands (channels) in a hyperspectral FC system is larger than the number of fluorochromes used. Therefore the mixing matrix $\mathbf{M}$ is not square and it cannot be inverted.

Traditionally, the approach used in imaging applications utilizes a least-square technique (LS), which is equivalent to computation of Moore–Penrose pseudoinverse. This same method can be applied to FC data:

$$\hat{\mathbf{a}} = \arg\min_{\mathbf{a}\in\mathbb{R}} \left\{ (\mathbf{r} - \mathbf{Ma})^T (\mathbf{r} - \mathbf{Ma}) \right\} \quad \Rightarrow \quad \hat{\mathbf{a}} = \left(\mathbf{M}^T\mathbf{M}\right)^{-1}\mathbf{M}^T\mathbf{r} \qquad (3)$$

Unmixing the flow cytometry data using LS may result in negative abundances (i.e., unmixed signal lower than zero). As in imaging, this problem can be solved by imposing physical constraints on the unmixing process: a non-negativity constraint assures that all the results are positive, while the sum-to-one constraint states that all the unmixed values must sum to 100 % of the mixed input fluorescence signal.

$$\sum_{i=1}^{p} a_i = \mathbf{1}^T\mathbf{a} = \sum_{i=1}^{p} r_i = \mathbf{1}^T\mathbf{r}, \quad a_i \geq 0$$

The minimization of $\|\mathbf{r} - \mathbf{M}\hat{\mathbf{a}}\|_2$ with the additional constraints above must be performed using numeric methods, as there is no closed-form solution to constrained minimization.

The simple unmixing models, in the unconstrained version (Eq. 3) or with the additional constraints above, work relatively well in typical imaging applications, and they may be applied to spectral FC data without any further modifications. However, it must be pointed out that the described model implicitly assumed that noise in Eq. 2 is an additive, Gaussian noise. Consequently, the ordinary LS solution operates under an assumption of data homoskedasticity (homogeneity of variance). However, the noise model in fluorescence observation is actually Poisson-like. Therefore the variance of the fluorescence intensity measurements increases with the intensity, resulting in heteroskedastic data.

This information can be incorporated in the unmixing model, as recently demonstrated in a report by Novo, Grégori, and Rajwa (Novo et al. 2013). Starting from the assumption of shot noise—limited measurements, the authors showed that the heteroskedasticity can be accounted for in unmixing by utilizing Poisson regression process defined within generalized linear models:

$$\widehat{\mathbf{a}} = \underset{\mathbf{a}\in\mathbb{R}_{\geq 0}}{arg\,min}\left\{2\mathbf{j}^T\left(\mathbf{r} \circ log\left(\mathbf{r} \circ \overline{\mathbf{Ma}}\right) - (\mathbf{r} - \mathbf{Ma})\right)\right\}$$

where $\mathbf{j}$ is an $L \times 1$ sum vector of 1, $\overline{\mathbf{Ma}}$ is a Hadamard inversion of $\mathbf{Ma}$, and $\bigcirc$ denotes element-wise multiplication (Hadamard product).

The methodologies described above assume that the mixing matrix $\mathbf{M}$ is known and does not change. However, the spectrum of fluorochromes is not stable and may vary owing to multiple experimental factors. Therefore, in some experimental settings both $\mathbf{a}$ and $\mathbf{M}$ should be treated as unknowns, but this leads to the problem of decomposing the measured signals into two non-negative matrices. This issue, known as non-negative matrix factorization, has been widely studied in the context of blind source separation (Bilgin et al. 2012; Lee and Seung 1999; Pauca et al. 2006; Rabinovich et al. 2003).

In their data acquisition/processing software Sony implemented an algorithm described by Sekino et al. that estimates not only the fluorochrome abundances $\mathbf{a}$ but also the mixing matrix $\mathbf{M}$ based on probabilistic modeling and Bayesian estimation.[3] In this model, the spectrum of each fluorochrome is assumed to be generated from a prior distribution that corresponds to our belief regarding the likely shape of the spectrum. For example, one may use the truncated normal distribution to model the prior:

$$\mathbf{m}_i \sim N_{m_i \geq 0}(\boldsymbol{\mu}_i \boldsymbol{\Sigma}_i)$$

[3] Masashi Sekino, Yasunobu Kato, Tatsumi Ito, "Probabilistic Spectrum Compensation for Flow Cytometry." Presented at CYTO2012, XXVII Congress of International Society for Advancement of Cytometry, 23–27 June, 2012, Leipzig, Germany,.

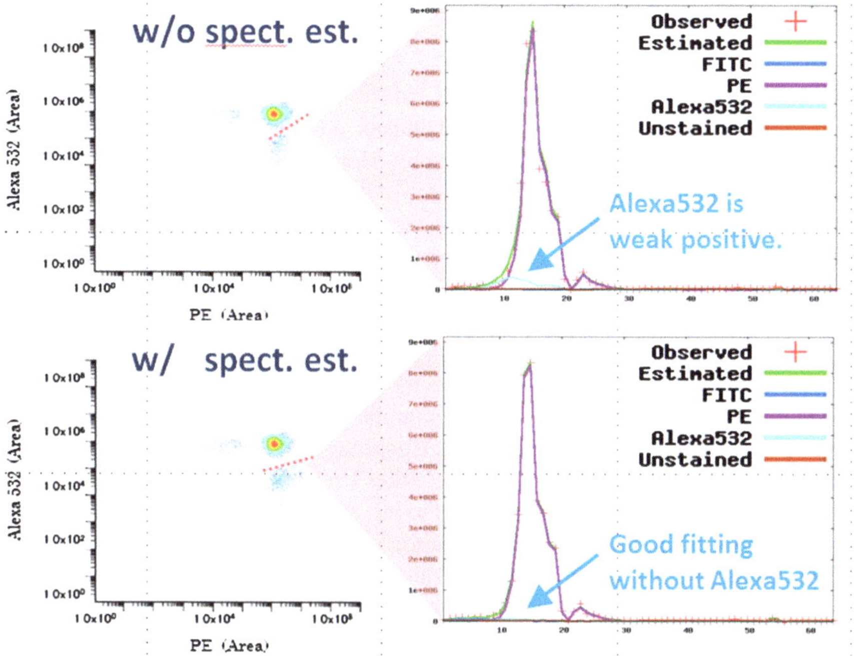


**Fig. 8** Dynamic estimation of the spectrum removes the artifact showing the presence of weak positive Alexa532 signal for cells labeled only with PE. After activation of the spectrum estimation technique the Alexa532 signal was correctly found to be approximately zero

Therefore, $\mathbf{m}_i$ is normally distributed with mean vector $\boldsymbol{\mu}_i$ and the covariance matrix $\boldsymbol{\Sigma}_{i,}$. These parameters can be estimated from data collected on samples stained with a single label. Obviously $\mathbf{m}_i$ should have only non-negative values. The measured signal is modeled by Eq. 2, with normally distributed error and the non-negativity constraint:

$$\mathbf{r} = \mathbf{M}\mathbf{a} + \mathbf{n}, \quad \mathbf{n} \sim N(0, \Lambda), \quad a_i \geq 0$$

The variational Bayesian algorithm enables an effective estimation of both **M** and **a** under these settings and constraints. Figure 8 demonstrates that the simultaneous estimation of **M** improves chances of obtaining a relevant estimation of **a**.

Finally, in some applications the multiplexing by spectral tags may not require spectral unmixing at all. In this setting it may be beneficial to classify the spectra directly, as opposed to classification based upon unmixed intensities. A large number of techniques may be utilized here, including unsupervised data reduction (using, for example, principal component analysis, independent component analysis, or factor analysis) or supervised techniques (such as neural networks or support vector machines) as illustrated in Fig. 5.

## 7 Potential Applications

The technology of spectral cytometry is still in its infancy; therefore, it would be premature to speculate on its potential impact on single-cell analysis or on the breadth and range of possible applications. However, it has already been demonstrated that the use of a spectral approach improves the ability to measure FRET-based molecular beacons used to identify specific stem-cell differentiation states (Bernstein and Hyun 2012). Generally, spectral analysis in cytometry is expected to improve the quality of all FRET-type measurements as has been shown in spectral microscopy applications (Zimmermann 2005; Zimmermann et al. 2002).

The number of bands collected in spectral cytometry experiments is larger than the number of fluorochromes used. This implies that spectral unmixing is always performed in an overdetermined setting. Consequently, accessing the spectral data one can unmix autofluorescence or any other background signal not originating from the labels of interest. This will lead to increases in measurement accuracy. As shown with spectral microscopy, acquisition of an entire spectral range allows for better separation of spectrally similar fluorescent proteins (Haraguchi et al. 2002; Zimmermann 2005; Zimmermann et al. 2003). Additionally, the presence of spectral information makes it possible to construct sophisticated unmixing strategies that take advantage of signal-formation models such as the recently proposed Poisson unmixing (Novo et al. 2013).

Finally, the use of spectral methodology can be advantageous if the spectra of the used labels change during the experiment. Mathematical techniques based on non-negative matrix factorization are capable of estimating both the label signals and the spectral matrices. Again, work developed for spectral imaging applications could be applied to spectral cytometry (Bilgin et al. 2012; Lee and Seung 1999; Pauca et al. 2006; Rabinovich et al. 2003).

## References

Asbury CL, Esposito R, Farmer C, van den Engh G (1996) Fluorescence spectra of DNA dyes measured in a flow cytometer. Cytometry 24:234–242. doi:10.1002/(SICI)1097-0320(19960701)24:3<234:AID-CYTO6>3.0.CO;2-H

Bagwell CB, Adams EG (1993) Fluorescence spectral overlap compensation for any number of flow cytometry parameters. Ann N Y Acad Sci 677:167–184. doi:10.1111/j.1749-6632.1993.tb38775.x

Bernstein HS, Hyun WC (2012) Strategies for enrichment and selection of stem cell-derived tissue precursors. Stem Cell Res Ther 3:17. doi:10.1186/scrt108

Bilgin CC, Rittscher J, Filkins R, Can A (2012) Digitally adjusting chromogenic dye proportions in brightfield microscopy images. J Microsc 245:319–330. doi:10.1111/j.1365-2818.2011.03579.x

Brown LO, Doorn SK (2008a) A controlled and reproducible pathway to dye-tagged, encapsulated silver nanoparticles as substrates for SERS multiplexing. Langmuir ACS J Surf Colloids 24:2277–2280. doi:10.1021/la703853e

Brown LO, Doorn SK (2008b) Optimization of the preparation of glass-coated, dye-tagged metal nanoparticles as SERS substrates. Langmuir ACS J Surf Colloids 24:2178–2185. doi:10.1021/la703218f

Buican TN (1990) Real-time fourier transform spectrometry for fluorescence imaging and flow cytometry. In: Proceedings of SPIE 1205. SPIE, Los Angeles, CA, pp 126–133

Garini Y, Young IT, McNamara G (2006) Spectral imaging: principles and applications. Cytometry A 69A:735–747. doi:10.1002/cyto.a.20311

Gauci MR, Vesey G, Narai J et al (1996) Observation of single-cell fluorescence spectra in laser flow cytometry. Cytometry 25:388–393. doi:10.1002/(SICI)1097-0320(19961201)25:4<388:AID-CYTO11>3.0.CO;2-R

Goddard G, Martin JC, Naivar M et al (2006) Single particle high resolution spectral analysis flow cytometry. Cytometry A 69A:842–851. doi:10.1002/cyto.a.20320

Grégori G, Patsekin V, Rajwa B et al (2012) Hyperspectral cytometry at the single-cell level using a 32-channel photodetector. Cytometry A 81A:35–44. doi:10.1002/cyto.a.21120

Haraguchi T, Shimi T, Koujin T et al (2002) Spectral imaging fluorescence microscopy. Genes Cells Devoted Mol Cell Mech 7:881–887

Johnson PE, Lund ML, Shorthill RW et al (2001) Real time biodetection of individual pathogenic microorganisms in food and water. Biomed Sci Instrum 37:191–196

Keshava N, Mustard JF (2002) Spectral unmixing. Sigl Process Mag IEEE 19:44–57. doi:10.1109/79.974727

Lee DD, Seung HS (1999) Learning the parts of objects by non-negative matrix factorization. Nature 401:788–791. doi:10.1038/44565

Loken MR, Parks DR, Herzenberg LA (1977) Two-color immunofluorescence using a fluorescence-activated cell sorter. J Histochem Cytochem 25:899–907. doi:10.1177/25.7.330738

Nielsen AA (2001) Spectral mixture analysis: linear and semi-parametric full and iterated partial unmixing in multi- and hyperspectral image data. J Math Imaging Vis 15:17–37. doi:10.1023/A:1011269530293

Nolan JP, Condello D, Duggan E et al (2013) Visible and near infrared fluorescence spectral flow cytometry. Cytometry A 83A:253–264. doi:10.1002/cyto.a.22241

Novo D, Grégori G, Rajwa B (2013) Generalized unmixing model for multispectral flow cytometry utilizing nonsquare compensation matrices. Cytom Part J Int Soc Anal Cytol 83:508–520. doi:10.1002/cyto.a.22272

Pauca VP, Piper J, Plemmons RJ (2006) Nonnegative matrix factorization for spectral data analysis. Linear Algebra Its Appl 416:29–47. doi:16/j.laa.2005.06.025

Perfetto SP, Chattopadhyay PK, Roederer M (2004) Seventeen-colour flow cytometry: unravelling the immune system. Nat Rev Immunol 4:648–655. doi:10.1038/nri1416

Rabinovich A, Agarwal S, Laris CA, et al (2003) Unsupervised color decomposition of histologically stained tissue samples. Adv Neural Inf Process Syst 16:667–674

Robinson JP (2004) Multispectral cytometry: the next generation. Biophotonics Int 11:36–40

Robinson JP, Rajwa B, Gregori G et al (2005) Multispectral cytometry of single bio-particles using a 32-channel detector. Proc SPIE 5692:359–365

Robinson JP, Rajwa B, Grégori G, et al (2007) Multi-spectral detector and analysis system. US Patent 7280204 B2

Roederer M (2001) Spectral compensation for flow cytometry: Visualization artifacts, limitations, and caveats. Cytometry 45:194–205

Roederer M, De Rosa S, Gerstein R et al (1997) 8 color, 10-parameter flow cytometry to elucidate complex leukocyte heterogeneity. Cytometry 29:328–339

De Rosa SC, Herzenberg LA, Herzenberg LA, Roederer M (2001) 11-color, 13-parameter flow cytometry: identification of human naive T cells by phenotype, function, and T-cell receptor diversity. Nat Med 7:245–248. doi:10.1038/84701

Settle JJ, Drake NA (1993) Linear mixing and the estimation of ground cover proportions. Int J Remote Sens 14:1159. doi:10.1080/01431169308904402
Shapiro HM (2003) Practical Flow Cytometry, 4th edn. Wiley-Liss, Hoboken, NJ, USA
Steen HB (1986) Simultaneous separate detection of low angle and large angle light scattering in an arc lamp-based flow cytometer. Cytometry 7:445–449. doi:10.1002/cyto.990070509
Stokke T, Steen HB (1986) Fluorescence spectra of Hoechst 33258 bound to chromatin. Biochim Biophys Acta BBA - Gene Struct Expr 868:17–23. doi:10.1016/0167-4781(86)90081-3
Wade CG, Rhyne RH Jr, Woodruff WH et al (1979) Spectra of cells in flow cytometry using a vidicon detector. J Histochem Cytochem Off J Histochem Soc 27:1049–1052
Wang J-CE, Kobie JJ, Zhang L et al (2009) An 11-color flow cytometric assay for identifying, phenotyping, and assessing endocytic ability of peripheral blood dendritic cell subsets in a single platform. J Immunol Methods 341:106–116. doi:10.1016/j.jim.2008.11.002
Watson DA, Brown LO, Gaskill DF et al (2008) A flow cytometer for the measurement of Raman spectra. Cytometry A 73:119–128. doi:10.1002/cyto.a.20520
Zimmermann T (2005) Spectral imaging and linear unmixing in light microscopy. Microsc. Tech. Springer Verlag, Berlin/Heidelberg, pp 245–265
Zimmermann T, Rietdorf J, Girod A et al (2002) Spectral imaging and linear un-mixing enables improved FRET efficiency with a novel GFP2-YFP FRET pair. FEBS Lett 531:245–249
Zimmermann T, Rietdorf J, Pepperkok R (2003) Spectral imaging and its applications in live cell microscopy. FEBS Lett 546:87–92. doi:10.1016/S0014-5793(03)00521-0

# Index




Current Topics in Microbiology and Immunology (2014) 377: 211–215
DOI: 10.1007/978-3-642-54827-7

MIX
Papier aus verantwortungsvollen Quellen
Paper from responsible sources
FSC® C105338
FSC
www.fsc.org

If you have any concerns about our products,
you can contact us on
**ProductSafety@springernature.com**

In case Publisher is established outside the EU,
the EU authorized representative is:
**Springer Nature Customer Service Center GmbH**
**Europaplatz 3, 69115 Heidelberg, Germany**

Printed by Libri Plureos GmbH
in Hamburg, Germany